MACHINE VISION AND DIGITAL IMAGE PROCESSING FUNDAMENTALS

Louis J. Galbiati, Jr.
State University of New York
Institute of Technology

Prentice Hall, Englewood Cliffs, New Jersey 07632

Library of Congress Cataloging-in-Publication Data

Galbiati, Louis J.
    Machine vision and digital image processing fundamentals / Louis
J. Galbiati.
        p.   cm.
    Includes bibliographies and index.
    ISBN 0-13-542044-X
    1. Computer vision. 2. Image processing--Digital techniques.
I. Title.
TA1632.G35   1990
621.36'7--dc20   89-16038
                                                           CIP

Editorial/production supervision
  and interior design: **Karen Bernhaut**
Manufacturing buyer: **Denise Duggan**

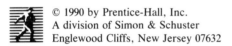 © 1990 by Prentice-Hall, Inc.
A division of Simon & Schuster
Englewood Cliffs, New Jersey 07632

The publisher offers discounts on this book when ordered
in bulk quantities. For more information, write:

    Special Sales/College Marketing
    Prentice-Hall, Inc.
    College Technical and Reference Division
    Englewood Cliffs, NJ 07632

Printed in the United States of America

10   9   8   7   6   5   4   3   2   1

ISBN   0-13-542044-X

Prentice-Hall International (UK) Limited, *London*
Prentice-Hall of Australia Pty. Limited, *Sydney*
Prentice-Hall Canada Inc., *Toronto*
Prentice-Hall Hispanoamericana, S.A., *Mexico*
Prentice-Hall of India Private Limited, *New Delhi*
Prentice-Hall of Japan, Inc., *Tokyo*
Simon & Schuster Asia Pte. Ltd., *Singapore*
Editora Prentice-Hall do Brasil, Ltda., *Rio de Janeiro*

CONTENTS

FOREWORD

The 1986 revenues for machine vision systems and components exceeded $400 million. By 1991 this figure will grow to over $2 billion. The **forecasted 40% compound annual growth rate** (CAGR) for Machine Vision systems will surpass that of automated manufacturing. Applications that will improve productivity using MV include product assembly, inspection, and material handling.

This **growth is forecasted despite apprehension** on the part of potential users. This apprehension is present, to some degree, in all but the leading edge members of each end using industry. It stems from a **lack of knowledge** of how to justify and to apply machine vision to their manufacturing operations.

The current number of total machine vision installations is small compared to the number of potential installations. Vendors and systems integrators agree that, outside of a few dozen in-house developments, the knowledge among end users is extremely limited. While the general concept is known, there is scant application information available for users to examine and relate to their own manufacturing operation.

PREFACE

This book is designed to provide the fundamental knowledge necessary to understand machine vision and digital imaging technology. The broad coverage of all aspects of machine vision as a system and the review of basic principals of optics, coupled with the building block approach, enables a reader with no previous knowledge to master the subject.

It is intended to provide individuals in industry with an overview of all facets of machine vision, to enable them to apply the technology to industrial applications. It includes material not usually found in a basic book on machine vision and covers the various aspects of vision technology that will be encountered in the application of machine vision and digital imaging to an industrial process.

The three to five principal machine vision concepts introduced in each chapter are in the order that they would be encountered in designing a total system. The basic concepts of image processing are covered to the extent that the individual will understand the techniques used in commercial vision systems in industry. Numerical examples are provided to help the reader understand the concepts and as a guide for individuals in industry in applying the concepts to specific applications.

The book is organized in a manner that it can be used for college students in an introductory course. It is based on lecture notes from courses, Robotic Vision in Industrial Applications, Robotic Vision and Digital Imaging Applications, and Special-Vision-Topics Seminar for students in electrical, industrial-manufacturing, mechanical engineering technology and the computer science programs of study at the SUNY College of Technology.

The references included at the end of each chapter provide an information resource for individuals who wish to obtain additional insight in the topics. Problems are included at the end of each chapter to provide practice in the application of the specific part of the technology and to assist the individual in determining how well the knowledge relative to the local objective has been mastered.

A chapter on bar coding has been included, since this technique is a specialized application of machine vision technology and is being widely accepted in the manufacturing industry. Bar coding can serve as a practical and cost effective mechanism for the introduction of vision technology in the industrial plant to increase productivity and for data acquisition in support of computer-integrated-manufacturing (CIM) systems.

ACKNOWLEDGMENTS

Special acknowledgment is made to the students in my classes for their critique of the material, and to J. Fred Johnson of the Union Fork and Hoe Company, Sam Krammer of the National Bureau of Standards, Peter D. Hiscocks and Brian Whelton of the Center for Advanced Technology at Ryerson University, Atlas Hsie of the Industrial Technology Program at SUNY College of Technology, and Congressman Sherwood Boehlert of the U.S. House Committee on Science and Technology for their technical input, encouragement, and assistance to me in initiating the machine vision research activity at the SUNY College of Technology. Appreciation is expressed to Irene Fluty and Sophie Trott for typing the manuscript and to the print shop staff for all the service they provided.

The cooperation and assistance is acknowledged of Automatic Identification Manufacturers Inc. (AIM), E. G. & G. Reticon, Itran Machine Vision Corp., Galileo Electro-Optics Corp., Newport Corp., Penn Video Corp., Photographic Sciences Corp., Photo Research, Poynting Products Inc., and Welch Allyn for permitting me to use and include their material in the book to make it relevant to the industrial world.

Louis J. Galbiati, Jr.

1

INTRODUCTION
TO MACHINE VISION
AND DIGITAL IMAGING

1.0 OVERVIEW

Machine vision and digital imaging technology is multidiscipline in the sense
that the field uses the knowledge of traditional engineering and computer
programing for the different parts of the process. The process can be subdivided into the following three activities:

1. Obtaining the digital representation of an image,
2. Employing computational techniques to process or modify the image
   data, and
3. Analyzing and using the results of the processing for the purpose of
   guiding robots or controlling automated equipment, assuring a level of
   quality in a manufacturing process, or supporting statistical analysis in a
   computer-assisted-manufacturing (CAM) system.

Individuals should have an understanding of the fundamentals of the entire
field before specializing in one area of the technology, since the field is very
broad and the different specializations impact on each other. Machine vision
and digital imaging will be a major field of endeavor for professionals in the
years ahead.

There has been fragmented use of vision technology during the past
three decades for space, military, and limited industrial applications. Machine vision was not used to a great extent because of the newness of the

technology: There was a lack of low-cost, commercially available equipment and a limited supply of individuals with technical knowledge about machine vision. The cameras and sensors were custom-made to support a specific task; and the Very Large Scale Integration (VLSI) process had not been developed to the point whereby it could be used to produce high resolution solid state sensors. The software had to be developed for each sensor and mission of the sponsoring project, as computational power was limited and costly. In addition, the application engineers involved in the projects had to learn on the job because there was little or no formal classroom instruction available on vision technology at the undergraduate level in our universities.

Digital image or machine vision technology will have a major impact on all industrial tasks in the next decade because the supporting technologies have progressed to the point where the use of this technology is now viable.

Three main conditions necessary for widespread application of a new technology are

1. Reliable hardware at reasonable cost,
2. Individuals who have the hardware and programming knowledge to apply the technology, and
3. A need or a problem requiring a solution.

All three of these conditions prevail today. Solid state detectors and personal computers provide relatively low-cost reliable sensing and computational capability. University educational systems are providing engineers with real-time programming skills in sufficient numbers, and there is a national need to increase productivity and quality in the United States if our standard of living is to be preserved. In addition, the change in the product liability climate is increasing the need for traceability in the manufacturing process and automatic data input to the product data base.

Widespread use of robotics and CAD/CAM in the industrial sector provides a need for an automated process of acquiring vision information in digital form. At the same time, social, industrial, and economic changes mandate a need to increase productivity in the manufacturing sector of the economy, improve levels of quality and reliability in the finished product, and provide better traceability of parts going into the item being manufactured for product-defect liability aspects.

Manufacturing operations were developed using vision capability by including a man "in the loop." The availability of vision for all human activity in the industrial sector was taken for granted. The introduction of robots in our factories makes it necessary to automate the vision function, since man is being eliminated from the loop. In order to gain insight into the problems associated with applying machine vision, it is necessary to determine how human and machine vision capabilities relate to each other.

1.1 VISION AND FACTORY AUTOMATION

Principal functions which can be performed by machine vision systems fall into three main categories:

1. Control
2. Inspection
3. Data input

Control, in its simplest form, is related to the determination of position and generation of appropriate commands to a mechanism to make it initiate or perform an action. Guiding automated vehicles (AGVs) in a factory material delivery system, directing a welding torch along the desired edge, or selecting the specific area to be repainted by a painting robot are some examples of control functions performed by machine vision systems.

Machine vision inspection applications relate to the determination and/ or quantification of parameters such as mechanical dimensions of a part as well as shape, quality of surfaces, number of holes in a part, presence or absence of specific features. The metrology activity is identical to the measurements made in the past by a human inspector with gauges and jigs. The nonmetrology activity includes checking labels, verifying the color and finish of a part, and/or the visible detection of unwanted foreign substance in a food product.

The inspection function may include determining if electrical characteristics or properties of a product are correct, by having the machine vision system observe the output of an electrical measurement meter as part of its routine. However, for such applications, it may be simpler and more economical to perform the process with a microprocessor and hard instrumentation, if that is the only reason for the presence of a vision system.

Information on product quality attributes as well as material and production traceability can be placed in the CIM data base by a machine vision system. This method of data input is highly accurate and reliable since the human element is eliminated from the loop. At the same time, it is extremely economical since the data has been assembled as part of the inspection process. The complexity of the vision system will vary; it may be a bar code reader to provide part identification required for inventory control purposes or a conventional industrial vision system if it is necessary to measure quality control parameters for product liability purposes.

1.2 HUMAN VISION VS MACHINE VISION

The concept of human vision from an industrial automation application involves a very complex mechanism in that the vision function cannot be isolated. The human vision system is integrated and interdependent on various sensors. The degree of coupling is adaptive and dependent on the brain and on signals from other organs in the body.

In addition, complex feedback loops, adaptive responses, and different degrees of signal processing exist at many locations in the body. For example, the rods and cones in the eye are sensitive to specific characteristics of the radiation entering the eye. The fatigue of the individual, illness, previous training, and knowledge affect the performance of the human vision system in ways that are not easily quantified. The comparative analysis, therefore, will be centered on a functional basis to as great a degree as possible.

1.3 COMPARATIVE FUNCTION PARAMETERS

Machine vision will be compared to human vision as a total system on the basis of functional parameters which are applicable to industrial or manufacturing processes. Since the list can be very extensive, the analysis in this section will be limited to functions which are of paramount importance in industrial applications such as

Adaptability
Decision making
Quality of measurements
Two and three dimensional capability
Economical considerations

Enhanced capability, as contrasted to cost reduction achieved by replacing human operators, is the key element in using vision systems to improve productivity. Special attention should be given to the elements in the functional capabilities which permit the development of manufacturing processes not possible with human operators.

1.3.1 Adaptability

Adaptability is the capability of a system to automatically adjust or modify its operations according to environmental parameters to achieve a desired result. An example of adaptability is the ability of an individual to take a second look at an object in order to make decision if there is uncertainty about a detail due to poor seeing conditions.

Machine vision system capabilities are very rigid; they are established by the hardware and software in the system. The system will repeat the vision process with a high degree of certainty once the unit is set up. The systems are adaptable in the sense that they can be reconfigured to make a different set of measurements.

The human system is highly adaptable in that **images** from different

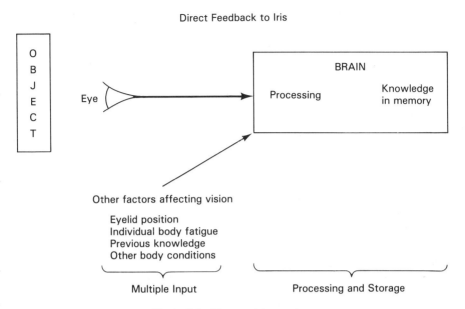

Figure 1-1   Human vision system.

angles will be used if there is an obstruction in the field of view, or images will be obtained from different ranges if a change of magnification is necessary to determine specific details. Human vision systems are dynamic in that the capability is determined by the physical characteristics of the eye as well as the intellect of the individual. This is illustrated in Figure 1-1. The intellect includes a number of items which cannot be quantified, such as learning, association, and perception. The human does this automatically and has the ability to learn by trial and error without reprogramming.

Requirements for most industrial manufacturing processes are very ridged and fixed as they are dictated by product specifications. Since manufacturing costs are directly affected by the "tightness of the requirements," it is desirable to make measurements to the necessary level of precision with a high degree of certainty. The nonadaptable characteristic of machine vision systems can be an advantage in industrial vision applications involving performance of identical processes on parallel manufacturing lines since it provides a high degree of measurement certainty.

In summary, the human vision system is much more adaptable than the machine vision system. The adaptability is the result of the human brain in the loop which can modify the system in an on-line fashion. Adaptability in a vision system is highly desirable in the development stage of process development where the values of the parameters to be measured have not been finalized.

1.3.2 Decision Making

At least one decision is required somewhere in the function performed by a vision system in an automated industrial application. The decision can be a factual judgement based on quantified measurable parameter data or it can be a value judgement based on morphologic, **heuristics,** and pseudoquantified factors.

The human vision systems can be effective in making value judgements for inspection tasks involving features, such as color, shape, odor, and the like, since it has perceptional and interpretational capabilities. However, the inclusion of the psychological and neurological inputs in the process can also lead to misinterpretations of results in other cases: For example, a line may appear longer than it is, depending on the features of the adjacent region of the image.

Machine vision systems require the quantification of the measurable parameters utilized in the decision making process. While the human vision system can base a decision on relative terms, like light or dark, the machine vision system requires a numerical value such as a specified **gray level** in a system having a given number of gray levels or the specific number of **pixels** between two reference points on the object. Pixels are the individual dot-like elements of the screen used to create an image.

Machine vision systems will be more consistent than human vision systems for factual-based decisions. However, a high level of technical expertise is required in defining the decision level of the parameters, and the parameters must be quantified. Human vision systems are easier to apply than machine vision systems for value-based decision applications. However, the consistency of the performance of the human systems is greatly impacted by fatigue, environmental factors, and the physical condition of the human.

1.3.3 Quality of Measurements

Consistency of results and *level of precision* are two main factors in the quality of measurements. Machine vision systems are clearly superior to human vision systems in the case of applications where the measurement is based on quantified input data.

The human vision system can discern something on the order of ten to twenty gray levels even though it is able to distinguish differences between many more levels on a comparative basis. Mechanical aids can be used to enhance the number of discernable levels. The performance of the human vision system will vary over time due to such factors as fatigue, environmental conditions, or distractions from other individuals in the work force.

The gray level capability of the machine vision system is related to and limited by the number of bits available to encode the integer representing the gray level. A four-bit system can have a maximum of 16 gray levels, since the largest number which can be represented by a four-bit binary number is 16. An eight-bit binary number must be used to represent 256 gray levels in a vision system. The 16- and 32-bit processors make it possible to increase the number of gray levels, but currently there are few, if any, industrial systems with more than 256 gray levels.

The machine vision system has no random errors due to human fatigue or distractions and the established level of performance will be constant for all practical purposes, up to the operating life of the equipment. Camera sensitivity is sufficient to provide signals for as many gray levels as needed.

1.3.4 Speed of Response

A machine vision system's image acquisition time depends on the size of the image matrix, the processing time of the **frame-grabbing** electronics, and the type of camera. Tube type cameras operating in a conventional **RS-170** mode will produce 30 images per second for standard commercially available monitors; the number of images per second could be increased by an estimated factor of five or ten by using a non-RS-170 mode. Solid state cameras can acquire the image in as little as ten microseconds (0.00001 seconds); the time required to read out the signal from the sensor will depend on the size of the matrix, **processing speed,** and system **bandwidth.** The use of parallel processing techniques can reduce the time by a factor equal to the number of parallel paths used.

The response time of the human vision system is on the order of 0.06 seconds or one sixteenth of a second. This is confirmed by the fact that we cannot detect the 30 image **frame** changes per second in the commercial TV monitors.

Machine vision systems are used in industry to inspect labels on bottles at a rate of 900 bottles per minute or about one bottle per 0.07 seconds. Higher rates are possible by increasing the number of bottle images in each frame. The human system task response time is on the order of one second or 60 tasks per minute maximum. The speed of response of the human system will be decreased by fatigue and environmental conditions.

In summary, image acquisition using machine vision systems is as much as ten times greater than that of the human vision system. This ratio is increasing with time as the state of the art in electronics improves, while that of the human system is not changing. The task response capability of the machine vision system is on the order of fifteen times greater than that of the human system.

1.3.5 Spectrum Response

The response of the human eye is narrow in that it can only use visible light (about a 400 millimicron span on the electromagnetic spectrum). The range is from violet at 390 millimicron to red at 790 millimicron **wavelength** as illustrated in Figure 1-2. The machine vision system response is relatively wide by comparison (about 100,000 millimicron); the range is from gamma and X rays, in the short wavelength region, to infrared, in the long wavelength region of the spectrum.

Machine vision systems have the capability of creating a new image based on the combination of information from different portions of the spectrum. For example, information in an infrared wavelength image can be used in conjunction with information in a visual wavelength image to identify the physical object acting as a source of heat energy in an industrial manufacturing process. The heat loss due to lack of insulation on a furnace or heat generated by friction in a bearing are examples encountered in industrial applications.

The color capability of the human eye is complex in that it does not separate the observed color radiation into basic components. Instead, it averages the energy at different wavelengths and identifies the color as that of an intermediate wavelength. The machine vision system color capability requires three data elements, one each at the red, green, and blue wavelengths. The production of color on a monitor is achieved by energizing the three components in a manner to produce the resultant color. Three times the data storage capacity is required to store a color image, and the R, G, and B components of a color image increases the amount of data to be processed over that for a monochromatic image.

In summary, the machine vision system has a much wider spectrum response than the human vision system. It is able to combine and use image information from different parts of the spectrum. The color capability of the machine vision system is more consistent and precise than that of the human eye.

1.3.6 Two and Three Dimensions

Three dimensional capability of the human vision system provides the capability of generating information on distance. This is taken for granted in everyday operations like guiding a vehicle or determining which of two objects is on top in a bin picking operation. Machine vision systems can be given three-dimensional (3-D) capability by using two or more cameras and complex processing techniques.

While two-dimensional capability is available in both human and machine vision systems, they differ in that the machine vision system data is quantified and can be used for making measurements to the degree of precision limited by the resolution of the array matrix. Human vision can make a fast estimate of the distance between two points on a surface, but a mechanical reference, such as a ruler, is required to provide precision.

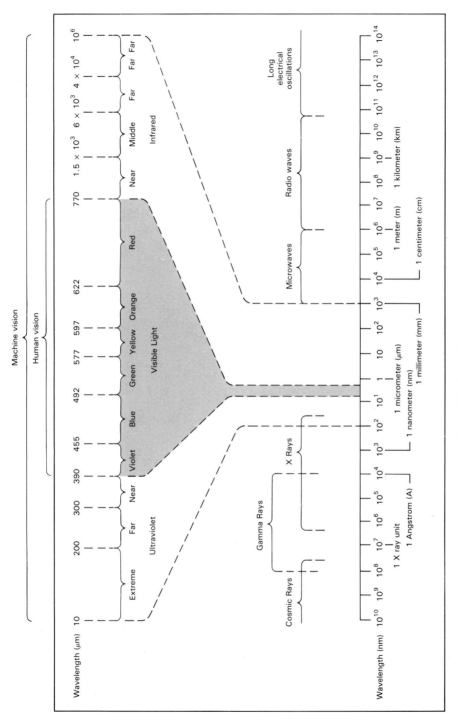

Figure 1-2 Image input spectrum.

TABLE 1-1 Comparison of Human and Machine Vision

	Human	Machine
Flexiblity	Very adaptable and flexible as to task and type of input.	Rigid as to the task; requires quantized data (pixels).
Ability	Can make relatively accurate *estimates* on subjective matters. Example: Detect bad fruit based on color, texture, shape, smell.	Can make dimensional measurements based on predetermined data inputs. Length of part or dimension of hole based on pixel count.
Color	Subjective in color.	Measures the magnitude of the chromatic parameter (R,G,B).
Sensitivity	Adaptive to lighting conditions, physical characteristics of item's surface and distance to object. Limited in the ability to distinguish between shades of grey. Varies as function of individual as well as from day to day. Can identify about 7 to 10 levels of gray.	Sensitive to level and frequency of illumination as well as physical nature of surface and distance to object. Ability to quantize is relatively high and fixed by sensor, environment and system characteristics. Can easily identify 256 levels of gray.
Response	Speed of response is slow and on the order of 1/10 second at best.	Speed of response is very high and dependent on system computation and bandwidth. Speeds of response on order of 1/1000 of second and higher are technically achievable.
2-D and 3-D	Can handle 3-D tasks and multiwavelength (color wavelengths) easily.	Can handle 2-D tasks easily but limited on 3-D tasks. Requires 2 cameras and is very slow in 3-D. Can perform mathematical operations on data, enhance data and use pixels for calculating areas, controls, and the like.
Data Output	Can manually supply secondary discrete data input. Cost is high and data subject to high error rate.	Can automatically supply precise discrete data input to the manufacturer's data base on a continuous basis at relatively low incremented cost basis.
Perception	Perceives brightness on logarithmic scale. Affected by surrounding area (background).	Can perceive brightness in either linear or logarithmic scale.
Spectrum	Limited to visual spectrum 300 to 700 millimicrons.	Can make measurements in entire spectrum from X rays to IR.

1.3.7 Summary of Comparison

Functions in the machine vision process complement each other; many are interdependent. A comparison summary of many functions in the application of machine vision is provided in Table 1-1. In general, all the human vision capabilities can be performed by machine vision systems, and the level of performance is superior to that of human vision from the standpoint of reliability and precision.

1.4 ECONOMIC CONSIDERATIONS

Implementation proposals for vision systems should include an analysis of the economic impact of the application. The economics of using machine vision for an industrial application include both a capabilities enhancement factor and a direct manufacturing productivity cost factor. Providing the capability to manufacture superior quality products by performing 100% inspection to insure that every item meets specifications is an example of improved capability factor. At the same time it can also result in increased productivity and reduced unit cost. The analysis of cost factors involved in replacing a human inspector with an automated vision inspection station will provide insight into the complexity of the problem as it is not always possible to quantify all factors. The determination of the cost of the impact on remaining employees of replacing a human operator with an automated station may not be easy to determine since it can involve one or more value judgements. The analysis should include consideration of this issue.

Example
Determine cost of replacing a human inspector with an automated vision system inspection station. The machine vision system in the example is leased rather than purchased to simplify the calculations relative to the per unit machine costs.

OPERATING DATA
1. The plant is operating one 40-hour shift per week at a production rate of 1000 finished objects/month.
2. Error rate is one defect per hour for human inspectors and zero defects per hour for machine vision stations.
3. Cost of correcting defective parts later in the production cycle is $10/part.
4. Vision inspection station leasing cost is $3000/month with an operating cost of $200/month.
5. Human inspector pay rate is $8/hour; fringe benefits rate is 24%; and plant overhead rate is 60%. Assume 172 working hours/month.
6. Individual to be replaced is 55 years old and has been with the company 17 years.
7. Sixty of the 500 nonunionized workers in the plant are in the quality assurance

function. The turnover rate of the plant is 12%/year and the sickness rate is 1%/year.

Solution   First calculate the cost of leasing and maintaining a machine on a per-part-inspected basis.

$$\text{Cost/part} = \frac{(\text{Leasing \& Operating Cost/month} \times 160\%)}{\text{Number of parts/month}}$$

$$= \frac{(3000 + 200) \times 1.60}{1000} = \$5.12/\text{part}$$

Next, to form a basis of comparison, calculate the cost of performing the same operation using a human inspector. This has two parts, labor costs and the costs of later correcting defective parts, a cost which is applicable with human inspectors only. Labor costs are

Base salary/hour	$8.00
Fringe rate (24%)	1.92
	$9.92
Overhead rate (60%)	5.95
Total labor cost	$15.87/hour = $2729.00/Month
Cost per part	$2729.00/1000 parts = $2.73/part

The costs related to defective parts are calculated as follows.

Direct correction cost per part	$10.00
Overhead (60%)	6.00
Correction cost per part	$16.00/defective part

The defect correction cost per month is computed by multiplying the number of defective parts by the correction cost per part.

172 defects/month × $16.00/defective part = $2752.00/month
Cost per part is $2752.00/1000 parts = $2.76/part

The human inspection cost is the sum of these.

$2.73 + $2.76 = $5.49/part

To find the direct savings cost, first subtract the machine vision cost from the human inspection. This gives the saving per part. Multiply this by the number of parts per year to get the saving per year.

Human inspection cost per part	$5.49
Machine vision inspection cost per part	5.12
Cost saving per part	0.37

Direct saving = Number of parts/year × saving/part
= 12,000 × $0.37 = $4,440/year

This completes the calculation of direct costs and savings. Next we must consider intangible costs. The individual being replaced by the inspection station can either be terminated or carried until an opening occurs. The estimated employee-termination cost of decreased morale, lack of employee cooperation to future plant upgrading activities, and the like should be included in the analysis. The out-of-pocket cost of carrying the employee until an opening occurs can be estimated with a high degree of certainty on a statistical basis from the operating data. From the following calculation we find that an opening can be expected in 7.2 weeks.

$$60 \text{ QA workers} \times 12\% \text{ turnover} = 7.2 \text{ terminations/year}$$
$$= 0.138 \text{ Terminations/week}$$
$$1 \text{ opening}/0.138 \text{ openings/week} = 7.2 \text{ weeks}$$

Next, we must calculate what it will cost to keep the employee during this period. For 24 hours a week the employee will fill in for workers absent because of sickness.

$$60 \text{ QA workers} \times 40\text{hr/week} \times \text{rate of sickness}$$
$$60 \times 40 \times 0.01 = 24 \text{ hours/week}$$

This involves no additional cost. For the additional 16 hours per week the employee will be considered to be working on undefined special projects. The cost for this is real and must be considered.

$$7.2 \text{ weeks} \times 16\text{hours/week} \times \$8.00/\text{hour} \times 1.24 = \$1143.00$$

The net saving in the first year will be the net saving previously calculated minus the special project employee cost.

$$\text{Net saving} = \text{Direct saving} - \text{Special project cost}$$
$$= \$4440.00 - \$1143.00 = \$3297.00/\text{year} = \$0.275/\text{part}$$

The saving over current cost during first year, exclusive of the impact on plant employees by the way the replaced employee is handled.

$$= \frac{\$0.275}{\$5.49} = 5.0\%$$

In subsequent years there will be no special project cost (that is for 7.2 weeks only) so the direct saving will be the previously calculated \$4440.00 per year or \$0.37 per part.

The morale effect could easily be in the one to two percent range, but will differ according to conditions at the plant. The impact of the benefits derived from the special project were not included to simplify calculations.

Changing wage scales will dramatically increase savings over the years since the bulk of the cost relative to the machine vision system is fixed.

Finally, economic considerations must include the impact of both the direct and the intangible costs. The importance of the nondirect cost is demonstrated by the example. The vision system installation would result in a 5% cost reduction with no employee being terminated.

The improvement in quality of product could result in improved sales with the result that production would have to be increased. The savings per unit will be greatly increased if the system is used on multiple shifts. The vision system can also perform a data input function to the Computer Integrated Manufacturing (CIM) system at essentially zero additional cost and an error rate of one error per five million input characters as compared to the error rate for a human operator manual input of one error per 300 input characters.

REFERENCES

1. Stan Lapidus and K. E. Garofano, "A Revised Look at Machine Vision Trends," *Manufacturing Systems,* March, 1987.
2. Eric Mittlelstadt, "Robotics and Vision: What Lies Ahead for Automation," *Robotics World,* January, 1987.
3. William Morling, *Hands That See,* Center for Automation and Intelligent Systems Research, Case Western Reserve University, Cleveland Ohio, 1986.
4. Larry Werth, "Automated Vision Sensing in Electronic Hardware," *Sensors,* December, 1986.
5. David J. Larin, "Vision's Next Steps," *Manufacturing Engineering,* December, 1986.
6. Lois Kane and Catherine Behringer, "Machine Vision in Europe," *Vision Technology,* November, 1986.
7. Cameron Geralds and Kenneth Pielowski, "Machine Vision as a Process Management Tool," *Vision-MVA/SME,* August, 1986.
8. Martin N. Levine, *Vision in Man and Machine* New York: McGraw Hill, 1985.
9. B. G. Batchelor, D. A. Hill, and E. D. C. Hodgson, *Automated Visual Inspection,* IFS Publications, Kempston, England, 1985.
10. James K. West, "Visual Fixturing for Robotic Assembly," *Robotics World,* July, 1985.
11. Alan Pugh, *Robot Vision,* IFS Publications, Kempston, England, 1983.
12. "The Advent of Machine Vision Systems," *Manufacturing Engineering,* November, 1982.

EXERCISES

1. A machine vision system can be leased for $500 per week and its operating cost is $20 per week. The vision system would be used for quality control inspection and displace one worker. The worker base rate is $8.00/hour; fringes are 24%, and the corporation overhead rate is 40%. The line is producing 40 pieces/hour; the machine vision system identifies two extra defective parts per hour, and the additional cost associated with replacing defective parts later in the manufacturing process is $1.25. Assume the line is running one 40-hour shift consisting of five 8-hour days.

   The employee that would be replaced by the vision system has 14 years of service and is 54 years old. The plant employs 350 people; 10% of the employees are in the quality assurance functions; the turnover rate is 6% per year and the sickness rate is 0.7%.

   Should the vision system be installed? Discuss the economic and other considerations relative to installing the vision system.
   (a) What are the relative or differential cost factors involved in using the machine vision system and the technical or engineering considerations on the application of the vision system?

(b) What are the managerial overall considerations?

(c) Would you change your recommendation if the worker were 32 years old with three years of service?

2. Use the conditions described in Problem 1, however, sales have increased to 1920 pieces per week. Should the vision system be installed? Perform a cost analysis and discuss the basis for your discussion.

3. A manufacturing line is producing 400 objects per hour; all objects must be checked to ensure that the label is correct. What are the considerations relative to using a human or a machine vision system.

4. Define an inspection system for checking
   (a) Foreign objects in baby food jars.
   (b) Foreign objects in one pound frozen fish packs.
   (c) That three parts have been inserted in an assembly where there are 25 different types of assemblies being produced.

5. Discuss the considerations of using machine vision and human vision for inspection where the environmental light level is changing by 40%.

6. Discuss the main considerations for the application of human vision, machine vision, or mechanical methods on the following applications:
   (a) Labels on bottles at 378/minute.
   (b) Four parts assembled on basic unit at 7 pieces/minute.
   (c) Foreign particles in baby food cans.

7. The cost elements on a plant interested in installing a vision system are as follows:

FICA = 7.2%;

Unemployment compensation: State 5%, Federal 1.5%

Workmen's compensation insurance: 4.8%

Medical insurance: $1000/year for full-time employees

Holidays: 4%, Vacation: 4%, Sick & personal time: 0.75%

Pension or retirement fund: 4%

The overhead burden rate is 60%

Determine the relative cost of a human operator vs a vision system if the turnkey vision system can be leased for 3%/month of the total cost of the system. The cost elements are

Basic vision system: $58,000

Tooling: $18,000

Special programming: $21,000

Installation: $9,000

Operating and servicing costs $300/month

Base rate of the worker is $12/hour.

Assume the machine does the work of one employee, that there are production needs for 2300 pieces/week and that the vision system detects one additional defective part/hour out of a production rate of 60 pieces/hour. The replacement cost of a defective part later in the process cycle is $9.00.

8. What are some of the main differences between human and machine vision which impact on the application of vision systems in the industrial environment.

9. What is a pixel and what values are associated with the pixel location in an array for a black and white image and a color image.

10. Describe some applications of a machine vision system for inspection and quality assurance tasks.

11. Discuss the types of industrial tasks that machine vision systems can be used for.

12. Discuss what parts of machine vision technology apply to your specific career goal.

13. Match the columns.

1. Machine vision digital imaging technology	_____	a. Smallest element of a scene over which the average brightness is determined.
2. Machine vision inspection function	_____	b. Measures the illumination in three portions (Red, Green, and Blue) of the spectrum.
3. Human vision system	_____	c. Measure of irradiance (brightness) given in integer values.
4. Machine vision system	_____	d. Highly adaptive capabilities, dynamic, and effective in value judgements.
5. Color system	_____	e. Determination and/or quantification of parameters such as mechanical dimensions, shape, or presence of special features.
6. Gray level	_____	f. Very rigid capabilities, high repeatability, and requires quantification of measurable parameters.
7. Pixel	_____	g. Can discern on the order of ten to twenty gray levels and distinguish between many more levels on a comparative basis.
8. Machine vision control function	_____	h. Obtains digital representation of an image, utilizes computational techniques to modify the

image data, analyzes and utilizes the results for various purposes.

9. Human sight   _____   i. Determination of position and generation of appropriate instruction signals.

2

WHAT IS
A VISION SYSTEM

2.0 SYSTEM OVERVIEW

A machine vision system is comprised of all the elements necessary to obtain a digital representation of a visual image, to modify the data, and to present the digital image data to the external world. The system may appear complex in an industrial environment due to all the associated manufacturing process equipment used in the application. The complexity is reduced when the system is viewed in terms of the three main functional components:

1. Image acquisition
2. Processing
3. Output or display

The majority of vision systems applications currently used in industry are in the areas of bar coding, desk-top publication, copy preparation for printing, and factory automation. The price range for a system varies from about ten thousand dollars for a PC version to over a million dollars for the complex systems used in the printing and auto industries.

Bar coding *commercial world* applications range from automatic data collection for inventory and production control to increasing productivity with point of sale automated scanning registers. The bar coding segment of the industry is highly standardized since the system performance requirements for commercial purposes are very similar for the stores selling various types of merchandise. The use of bar coding in the manufacturing sector is increasing rapidly due to the Department of Defense requirement that bar coding be on most items they purchase.

The equipment used in bar coding is simple but contains the three vision system components. Since the development activity of the vision aspects in the bar coding technology is very specialized, bar coding will be covered as a separate topic in a later chapter.

Applications in desk-top publication include the scanning capabilities of machine vision technology as a data input mechanism to digitize both printed text and photographs, the generation of special effects, and the preparation of material in a digital format for use by laser printers or display devices.

Factory automation applications involve the use of vision technology in the inspection tasks to improve the quality of the products produced, in the data collection tasks for inventory and management control, and in the process or machine control tasks for improving manufacturing productivity.

A simple industrial vision system used for factory automation could be characterized by a single camera monitoring an assembly line as shown in Figure 2-1. The vision system observes the object, determines if it is within specifications, and generates command signals according to the the determined results. The image acquisition equipment includes the lights, camera, and possibly the frame grabber. The processing equipment includes both

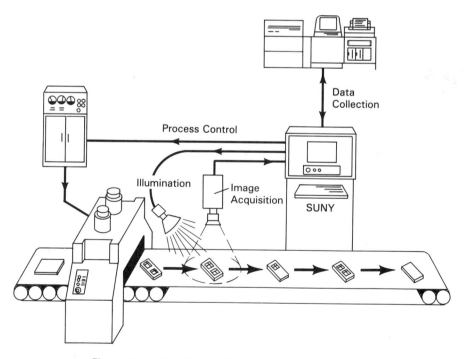

Figure 2-1   Industrial manufacturing cell with vision system.

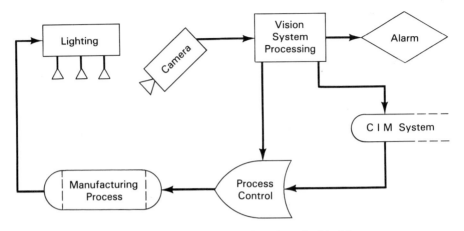

Figure 2-2   System functions in manufacturing cell with vision system.

hardware and software in the vision processing unit, and the output equipment is the electronics interfacing the system to various parts of the manufacturing world; examples are the process controller, CIM, and/or alarm.

The electrical instruction signals control the unit, taking objects off the assembly line and placing them in accept or reject containers, according to the quality. Data is transmitted to the computer-integrated-manufacturing system for statistical analysis and inventory control, which sounds an alarm if something is wrong.

The vision-manufacturing aspects of the assembly system are indicated in Figure 2-2. All functions or parts are highly interrelated, and an understanding of each part is important in achieving the objective of the application or task.

2.1 IMAGE ACQUISITION

Image acquisition transforms the visual image of a physical object and its **intrinsic characteristics** into a set of digitized data which can be used by the processing unit of the system. The acquisition function can be considered as consisting of four phases:

1. Illumination
2. Image formation or focusing
3. Image detection or sensing
4. Formatting camera output signal

2.1.1 Illumination

Illumination is a key parameter affecting the input to a machine vision system since it directly affects the quality of the input data and may require as much as 30% of the application effort. It is necessary to customize the illumination design for each application since there is essentially no standardized general purpose machine vision illumination equipment. The method and specific source of light energy affects the amount of processing and achievable results.

Many industrial machine vision systems in the past have used visible light since the sources were readily available and the application frequently was the automation of a manual inspection task. The inspection task was based on the capabilities of the human operator who worked within the visible portion of the spectrum. Four types of visible lamps most frequently used in the industrial environment are incandescent, fluorescent, mercury vapor, and sodium vapor. However, the use of illumination outside the visible spectrum, such as X rays, ultraviolet, and infrared is increasing due to the need to achieve special inspections not possible with visual light.

The methods for industrial applications can be subdivided into four categories:

1. Back lighting
2. Front lighting
3. Structured lighting
4. Strobe lighting

Environmental illumination affects all lighting methods by altering the total level of illumination on the object which manifests itself as noise in the data. The effect of environmental lighting can be be minimized by the use of light shields and barriers which prevent or reduce the amount of stray radiation entering the lens.

Conventional light sources are not always stable enough to insure a specific image quality. The light energy output of the source varies due to the age and the operating characteristics of the lamp. Fluorescent lamp output, for example, decreases as much as 15% during the first 100 hours and then continues to decrease but at a lower rate from day to day. The fluorescent lamps are brightest at about 105 degrees F and are sensitive to the applied voltage, but not to the extent that incandescent lamps are. Hence, variations in image quality can be caused by light-level irregularities from voltage variations resulting from activation of nearby equipment and changes of room temperature during the work.

Therefore, it is desirable to continuously sense the light level and modulate it on a real-time basis. Commercial units are available at relatively low cost to perform this function. The entire acquisition portion of the vision system should be checked on a regular basis to ensure that the system is operating within acceptable limits.

Back lighting.   **Back lighting** is when an object is located between the light source and the camera as illustrated in Figure 2-3. This results in the creation of a silhouette of the object by the light not intercepted by an opaque

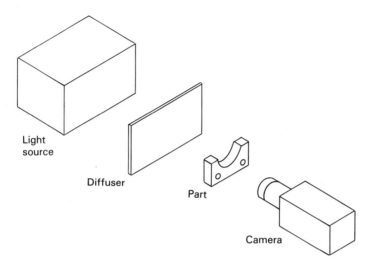

Figure 2-3   Back lighting: the light source is on the opposite side of the object from the camera.

object. A frosted glass is normally placed over the light source to produce a diffuse area emitter.

Back lighting has the advantage that it produces high-contrast images of the perimeter of opaque objects. The high contrast minimizes the image processing task and reduces the sensitivity of the system to illumination source variations. If the height of the object being inspected varies, it may be necessary to utilize a condensing system to produce a collimated light beam at the emitting surface.

The back lighting method is also ideally suited for tasks like the location of foreign material, voids, and fractures in transparent objects. The examination of bone fractures with x-ray plates and the measurement of heat energy leakage from a building with IR sensors are examples of this technique.

Back lighting can be combined with structured lighting by projecting a pattern on the background surface and measuring the difference between the object and the pattern. Information on surface characteristics, on features not visible in silhouette like the presence of bolts in blind hole, and on objects located on top of each other can not be obtained by this method of illumination.

The image in the back-lighted method is basically monochromatic in that a silhouette of the object is obtained. Since the edges of the object may not correspond to the pixel **boundaries** in the sensor, the edge pixels may have values between the minimum and maximum gray level values of the vision system. For example, the value of an edge pixel which is 50% covered would be 7 in a 16-level system; the value associated with each pixel on the edge will vary according to the portion of the pixel covered. The object in Figure 2-4 partially covers the area of some pixels and the gray level values

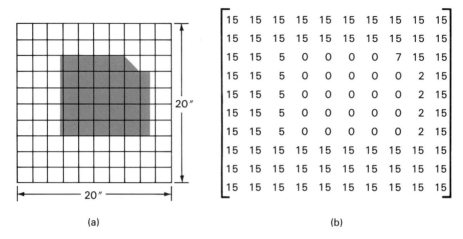

Figure 2-4 Back-lighted object (a) and image data (b).

obtained with a 16-level system are indicated in the image array. It should be noted that the reduced pixel value does not provide any information on the shape of the partial coverage, and a priori knowledge of the object shape must be used in conjunction with the measured image value.

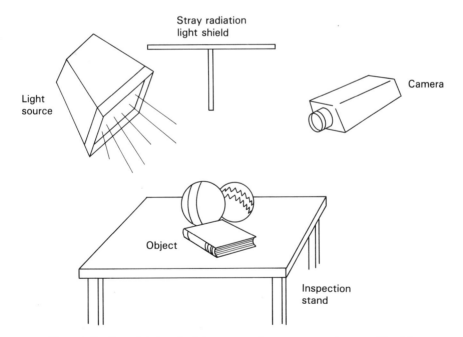

Figure 2-5 Front lighting: the light source and camera are on the same side of the object.

Front lighting.   **Front lighting** employs light reflected from the object. The illumination source and the camera are both on the same side of the object as illustrated by Figure 2-5. This method of illumination is used to obtain information on surface **texture** or features as well as for dimensioning. Either *specular* or *diffuse* illumination measurement techniques can be applied, depending on the angle of the camera.

Specular illumination measurement techniques are referred to as bright

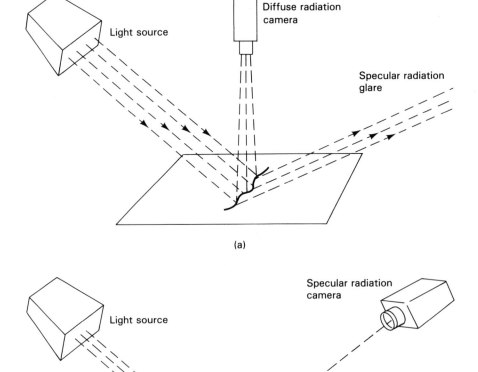

(a)

(b)

Figure 2-6   Specular and diffuse illumination. Defects show up as bright spots in a dark field in image obtained with camera utilizing diffuse radiation (a) defects show up as dark spots in a light field in image obtained with camera utilizing specular radiation (b).

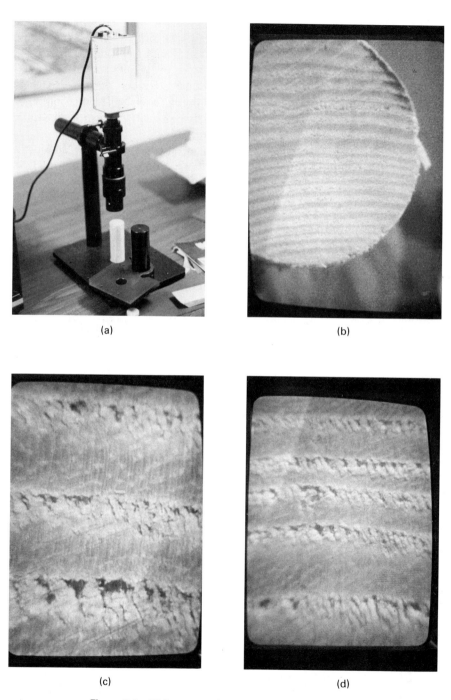

(a)   (b)

(c)   (d)

Figure 2-7   Vision system inspecting wood grain structure

field illumination; a smooth surface appears bright. The light rays from the illuminator are reflected directly from the object to the camera lens. Maximum energy reaches the imaging element if the camera is located at the same angular displacement to a perpendicular vector as the angular displacement of the light source. The reflected light intensity is very sensitive to angle changes of the object surface. The reflection of room lights on a computer monitor screen and the glare on a polished desk surface are undesirable examples of this type of illumination.

Diffused light measurement techniques are referred to as dark field illumination. A smooth surface appears black; light scattered from the surface is detected as illustrated in Figures 2-6 and 2-7. The camera is located at an angle perpendicular to the surface. A smooth surface will appear black except for the locations where there is a feature like a scratch, which scatters the light in many directions. Information on the object is provided by the scattered light sensed by the camera.

Strobe lighting. **Strobe lighting** is the illumination of the object by a short pulse of high-intensity light (5 to 500 microseconds duration) as illustrated in Figure 2-8. The short pulse of illumination can be used to freeze the motion of an object while the image is being acquired or to reduce the effect of adverse **ambient light.** The motion may be due to the object being on a conveyor belt or being part of a mechanism undergoing movement in a normal operating mode.

The light source and the camera must be synchronized since the pulse is of such short duration. The duration of the pulse is of importance to the extent that the object appears essentially stationary during the pulse. For example, an object ten centimeters long on a conveyor belt traveling at ten cm/second would have an image translation of 0.33 cm, or 3.3 pixels on a 100×100 array in the 0.033 seconds time period between adjacent RS-170 frames. The image will appear to be stationary since the object only moves a distance of 50 microcentimeters, or 0.005 pixels, during the 5 microsecond pulse duration.

The automatic iris adjustment must be disabled if a tube type camera is used. Otherwise, the camera sets the iris for the average light intensity and the camera will only detect the strobe illumination part of the cycle. The strobe light intensity should be at least ten times the ambient light.

If a solid state camera is used, an initiation signal must be supplied to the vertical framing circuit of the imaging camera. The image existing during the strobe illumination period is captured even though the data transfer to the frame-grabber unit may take considerably longer. Each solid state photodetector in the array retains its voltage level for the time period required by the camera system to sample all the detectors in the array.

Structured light. **Structured light** is the use of illumination of the object with a special pattern or grid. The intersection of the object and the projected illumination results in a unique pattern depending on the shape and

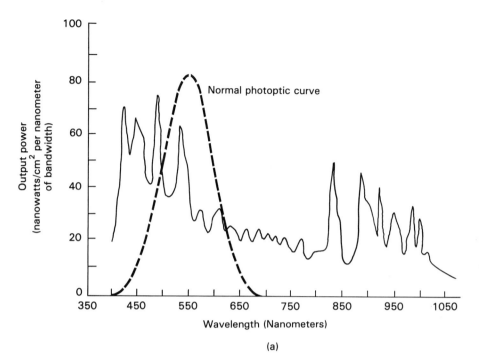

(a)

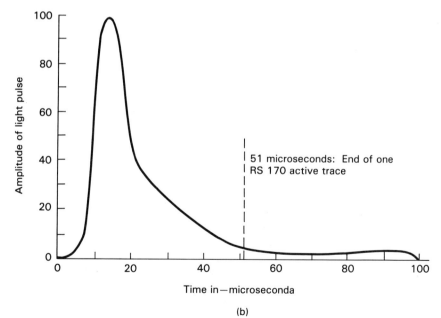

(b)

Figure 2-8   Strobe light pulse spectral distribution (a) and light output amplitude waveform (b).

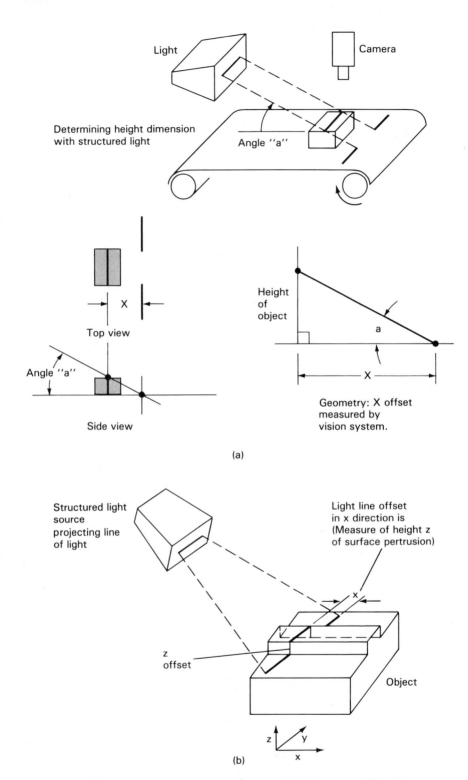

Figure 2-9 Structured light: geometry (a) angled structured light (b).

dimensions of the object as illustrated by Figure 2-9. A three-dimensional feature is converted into a two-dimensional image. Vertical and horizontal distances, as well as the shape of surface features, can be measured.

The line pattern in Figure 2-9 would be a straight line if there were no object present. The amount of displacement of the line segment on the top of the object is related to the height of the object for a given configuration. Additional information, like the distance between features, can be obtained by increasing the complexity of the pattern. The use of a circular light beam structure in spherical object inspection is shown in Figure 2-10 and a collimated offset angular light beam for inspecting parallel groves or holes is shown in Figure 2-11. The width of the dark lines in the image is a function of the angle of the light rays and the depth of the groves. The method of rounding off the **pixel value** must be known in order to determine tolerance values on the dimensions.

Generation of the structured light pattern can be accomplished at relatively low costs with a photographic negative in either a conventional projector or a photographic enlarger system. A flat structured light plane can be generated with a laser bar code scanner. The internal scanning of the illumination laser beam in the bar code reader unit projects a line on the surface of the object which can be detected by the vision system. Fiber optic bundles can be used to generate a precise light structure pattern that is remote

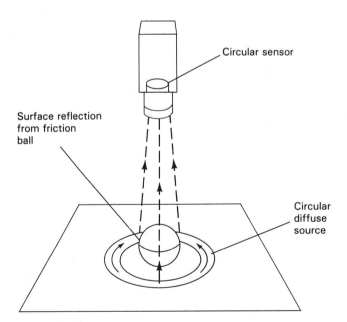

Figure 2-10   Circular structured lighting: sensor elements and illumination source are all equidistant from ball surface to obtain consistent defect signature as ball is rotated in elliptical fashion.

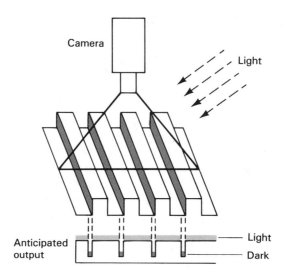

Figure 2-11   Offset structured lighting: shadows define features such as groves and holes. (Width of dark areas is a function of the angle of the light rays.)

from the light source and can be combined with the strobe technique to introduce time dimension into the pattern for detecting the rate of change of a surface.

2.1.2 Image Formation and Focusing

The image of the object is focused on the sensing element with a lens, in a way similar to that used in a photographic camera. The difference between the photographic camera and the machine vision system is that camera uses film, in contrast to a sensor in the machine vision system, to capture the image. The sensor converts the visual image to an electrical signal.

The machine vision camera is usually specified separately from the vision system as the capability and feature requirements are dependent on the applications. In addition, the camera lens parameters must be specified as it is the element which adapts the camera to the specific application. Four important parameters associated with optical lens of the vision system are

1. Magnification
2. Focal Length
3. Depth of Field
4. Lens Mounting

The optics associated with the lens parameters will be reviewed briefly in the following sections to provide a basis for the application of vision systems to industrial applications.

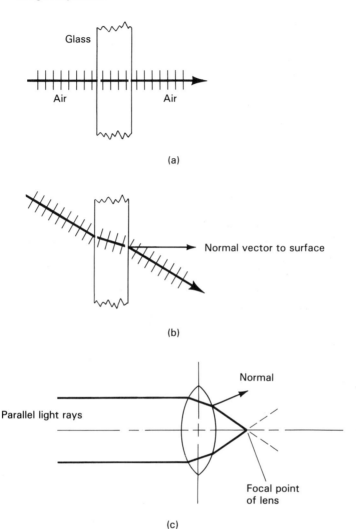

Figure 2-12   Path of light rays perpendicular to surface (a), at an angle to surface (b), and through a lens (c).

The focusing property of a lens is the result of light waves having a higher velocity of propagation in air than in glass or other optical material. Light rays crossing an air-to-glass junction will bend toward the normal vector of the interface surface as shown by the path of the light rays in Figure 2-12.

Rays exiting from a glass-air interface will be refracted away from the normal of the glass-air interface surface. Hence, a ray perpendicular to the air-glass-air interface will not be bent, and a wave entering the interface at an angle will exit at the same angle as the incident ray. It will be displaced by an amount that depends on the width and refractive index of the glass if the two surfaces are parallel as shown in Part b of the Figure 2-12.

A lens is an extension of the air-glass-air case where the two surfaces are spherical instead of being flat. All parallel light rays entering the lens will converge at a point known as the focal point. The location of the focal point is given by the expression

$$\frac{1}{f} = \frac{1}{R_1} + \frac{1}{R_2}$$

where f is the focal length distance from the center plane of the lens and R_1 and R_2 are the radius of the curvatures of two lens surfaces.

Rays from the object will continue in straight lines through the focal point and form an image on a plane located at any distance beyond the focal point. The image will be inverted and the size of the image will increase as the distance from the lens increases. The image detector or sensor array of the vision systems is located at a fixed position in this region of the optical system in the camera. The location of the detector is established by the manufacturer of the camera and hence is not a variable factor in the application.

Magnification.   Magnification (m) is a measure of the relative size of visual image of the object in the physical world to the size of the image formed on the sensor located at the detector plane in the camera. The value of m will be less than 1 in the case of conventional industrial applications, since the dimensions of the detector are smaller than those of the object being viewed. The value of m will be greater than one in the case of microscopic type applications.

The height of the image, H_i, in Figure 2-13 is a function of the distance between the lens and the location of the image plane in the camera. If the tip of the object arrow in Figure 2-13 is designated as o and the tip of the image arrow as i, the triangle (c,o,D_o) is similar to the triangle (c,i,D_i), and the ratios of corresponding sides are equal. Magnification is the ratio of image size to object size.

$$m = \frac{H_i}{H_o} = \frac{D_i}{D_o}$$

Where H_i is established by the size of the sensor array. D_i is the distance between the lens and the image plane; it is fixed when the camera is manufacturered. D_o is the distance between the object and the camera lens.

The system magnification specification is based on the maximum size of H_o (the size of the image that is to be projected on the sensor array or on D_o which is the distance between the lens and the object plane). The greater the distance D_o, the smaller the magnification.

$$\text{Magnification } m = \frac{H_i}{H_o}$$

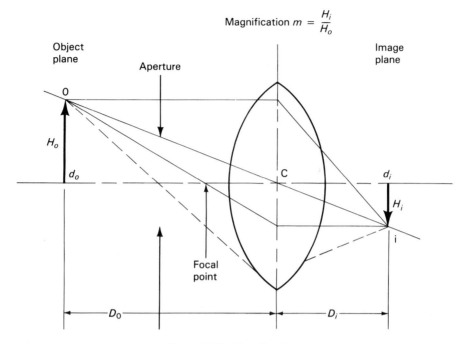

Figure 2-13 Magnification.

Example

Determine the magnification of the vision system and the size of a pixel on the sensor and on the object if the size of the 100 × 100 solid state sensor array is 0.15 × 0.15 inches and the size of the object to be measured is 3 by 3 inches.

Magnification

The magnification is the ratio of image to object size.

$$m = \frac{H_i}{H_o} = \frac{0.15''}{3.0} = 0.5''$$

Pixel size on array

The size of a pixel is equal to the distance between sensor elements. The pixel size is determined by dividing the dimension of the sensor array by the number of elements in the same direction.

$$\text{Pixel size} = \frac{\text{dimension of array}}{\text{number of elements}} = \frac{0.15''}{100} = 0.0015''$$

Pixel size on object

The size of the pixel on the object can be determined from the system magnification.

$$\text{Pixel size on object} = H_o = \frac{H_i}{m} = \frac{0.0015''}{0.05} = 0.03''$$

Focal length.   The focusing characteristics or focal length (f) of the simple lens is related to the curvatures of both surfaces of the lens by the expression

$$\frac{1}{f} = \frac{1}{R_1} + \frac{1}{R_2}$$

or

$$f = \frac{R_1 \times R_2}{R_1 + R_2}$$

where R_1 and R_2 are the radii of the curvatures of the lens surfaces. The focal length is related to two system parameters, magnification and the distance from the lens to the object, by the expression

$$f = \frac{D_o}{1 + 1/m}$$

Designation of a specific focal length lens establishes the distance between the camera and the object for a given magnification.

Example
Determine the focal length of the lens for a vision system with a magnification of 0.05 and a distance of thirty inches between the object and camera.
Solution

Focal length

$$FL = \frac{D_o}{1 + 1/m} = \frac{30''}{1 + 1/0.05} = 1.428''$$

Two dimension visual images located 30 inches from the lens will be in perfect focus on the image plane assuming the lens does not introduce aberrations. However, most (nonprinting) industrial applications involve three dimensional objects. The effect of a point of interest on the object not being on the object plane is to blur the image. That is, a point on the object will map into a two dimensional area on the image plane, as illustrated by Figure 2-14, since the rays diverge from a point at some distance different than the focal length. The relationship between the circle of confusion and the depth of field is illustrated in Figure 2-15.

Point a on the object plane maps into point a' on the image plane; point b, located between the object plane and the lens, will map into point b' at a location between the image plane and the lens. It will map into the area $b'' - b'$ known as the *circle of confusion,* on the image plane.

In a similar manner, point c, at a distance greater than that of the object plane, will map into a point c' at a distance greater that the distance of the

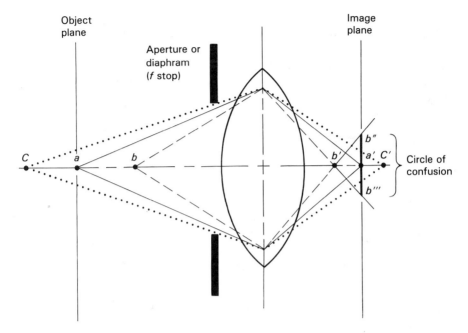

Figure 2-14   Lens system.

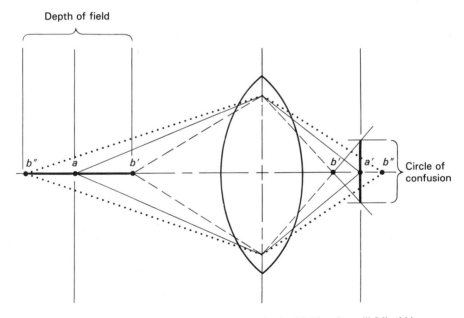

Figure 2-15   Depth of field. Any point in the depth of field region will fall within the circle of confusion on the image plane.

image plane to the lens. The image would appear blurred or out of focus in a conventional film camera.

The depth of field is that zone where defocusing will not adversely affect the quality of the measurement data. Every point of interest on a three-dimensional object located on the inspection surface must be in the depth-of-field zone as shown in Figure 2-16. The D_o dimension is taken from the center of the depth-of-field zone to the lens.

Depth of field.   The space above and below the object plane where the lens maintains the focus of the image within acceptable limits is the depth of field. The depth of field is of concern in vision technology when viewing objects located on a work surface where the point of interest is not located at the object plane. The depth of field is a function of aperture size, magnification, and size of the sensor elements. The depth increases as the aperture becomes smaller, but the amount of light transmitted decreases.

The adjustable aperture of the camera, used to vary the effective size of the lens opening is shown in Figure 2-14. An f stop of 16 represents the smallest aperture opening; the size of the opening increases as the f stop decreases. Each adjacent number on the descending order scale increases the relative amount of light transmitted by a factor of two. Standard f stops are 16, 11, 8, 5.6, 4, 2.8, 2, and 1.5.

Aperture size has two effects in the vision system: The smaller the lens opening

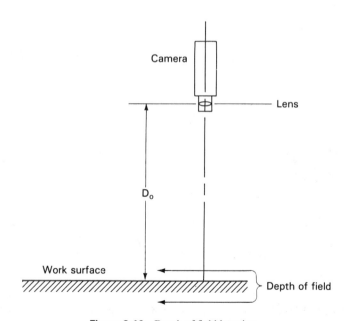

Figure 2-16 Depth of field location.

1. the greater the depth of field
2. the lower the amount of light transmitted to sensor

A high f-stop increases the height of the working range but requires an increase in the exposure time. The longer exposure time does not pose a problem for static images but does have an impact on measurement capability under dynamic conditions.

The smaller the pixel areas of the sensor, the smaller the depth of field. The relationship of the depth of field and the camera parameters is

$$\text{Depth of field} = \frac{2af(m + 1)}{(m)^2}$$

where a is the pixel size, f is the aperture size (f-stop of lens), and m is the magnification factor.

Example

Determine the depth of field for a vision system having a 200 × 200 array sensor 0.30 × 0.30 inches, an f-stop of 16, and a magnification factor of 0.05.

$$\text{Pixel size} = 0.30/200 = 0.0015''$$

$$\text{depth of field} = \frac{2 \times 0.0015 \times 16 \times (1 + 0.05)}{(0.05)\,(0.05)} = 20.16''$$

If the height of the object is less than one half the depth of field, the distance between the lens and the work surface can be used to determine the focal length of the lens. In applications having a magnification factor greater than one, the depth of field is relatively small. The depth of field for the application illustrated in Figure 2-17 limits the size of the surface irregularities which can be viewed.

The distance between the lens and work surface is 0.0368 inch with the f stop set at 16 to obtain maximum depth of field. The pixel size on the 100 × 100, 0.15 inch array sensor is 0.0015 inch. The magnification is 10.

$$\text{depth of field} = 2 \times 0.0015 \times 16\,(1 + 10)\,/\,(10 \times 10)$$

$$= 0.00528''$$

Surface irregularities equal to one half the depth of field, of 0.00264 inch can be imaged if the surface of the object is located at the image plane.

Lens mountings.   The type of lens mounting system must be specified when purchasing cameras and lens. The application determines the type of mount to a great extent. The four types of mounts normally encountered are

1. C-mount for machine vision,
2. U-mount for 35 mm cameras,
3. L-mount for fixed flat field installations, and
4. Bayonet mount for quick change 35 mm cameras.

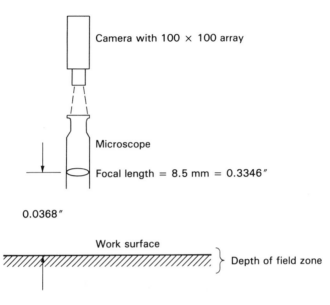

Camera with 100 × 100 array

Microscope

Focal length = 8.5 mm = 0.3346″

0.0368″

Work surface

Depth of field zone

Figure 2-17   Microscopic-type application.

The most common type of lens mounting in use today is the **C-mount;** some camera manufacturers are switching to the bayonet type mount on the camera and providing an adapter ring to accommodate the C-mount lens. C-mount lens systems originally were developed for commercial TV cameras. The size was designed for performance over the diagonal of a standard television vidicon tube.

The distance between the flange of the C-mount and the image plane focal distance point is 17.5 mm or 0.69 inch. The diameter of the usable area is such that array sizes up to 0.512 inch can be accommodated. It has a screw mounting with a one inch threading of 32 threads per inch.

The U-mount is widely used in 35 mm cameras. The distance between the flange and the image plane focal point is 47.52 mm or 1.79 inches (about three times longer than that of the C-mount). The diameter is about two and one-half times larger than that of the C-mount and can be used with arrays up to 1.25 inches in diameter. It has a screw mounting with a M-42 × 1 screw thread.

The L-mount is intended for fixed flat bed installations and has a M-39 × 1 screw mount. It can be used with a sensor array up to 1.25 inches in diameter.

The bayonet mount is designed to provide for easy lens changes on multilens cameras used with 35 mm cameras. The quick change feature is not required for machine vision systems but is used by some manufacturers for economic reasons. An adapter ring is required to put a C-mount lens on a camera with a bayonet mount.

2.1.3 Image Detection

Introduction.   Vision systems have an optical-electro device which converts electromagnetic radiation from the image of the physical object into an electrical signal used by the vision processing unit. The image is focused on the sensor in the camera by a lens. The sensor element located at the sensor plane in the camera produces an electrical signal representing the visible image. Cameras may have either a tube or a solid state sensor element.

Tube type cameras were originally developed for commercial television in the early 1930s, long before there were any solid state materials. They use a vacuum tube containing a light-sensitive element for image sensing; the internally captured image on the light-sensitive element is scanned to produce an analog voltage signal as output. Cameras connected directly to a conventional television monitor must have their output in RS-170 format.

Solid-state cameras have separate **photodiode** located at each pixel area as **detector** of the illumination from the object. The elements are usually arranged in either a **linear** array or a rectangular array. The solid state detectors were developed in the late 1960s as an outgrowth of the charge-coupled devices (CCD) invented at the Bell Telephone Laboratories. The output signal of the camera is obtained by sensing each diode in ordered sequence to obtain a series of voltage pulses representing the pixel value at the respective location. The voltage pulses must be converted into RS-170 format for viewing the image on the standard television monitor or digitized if it is desired to enter the data in computer memory.

Tube type cameras.   The five main types of image tubes used in vision systems are the newvicon, ultricon, plumbicon, silicon array, and saticon. The characteristics of each differ slightly as indicated in Table 2-1, and the selection is governed by the application.

Target element in the tube is a photosensitive surface where the image is formed by the lens. The charge density on the photosensitive surface builds up and is distributed in proportion to the intensity and duration of illumination flux impinging on the surface. The net result is that an electrical analog of

TABLE 2-1   Tube Material and Parameters

Type of Tube	Target Material	Dark Current nA	Sensitivity nA/lux	Lag %	Resolution L/P
Newvicon	Zinc selenide	1.0	250	12	500
Plumbicon	Lead oxide	0.8	40	3	750
Saticon	Selenium arsenic	1.0	15-30	2	750
Silicon Array	Silicon	0.5	200	0	240
Vidicon	Antimony-trisulfide	0.9	200	15	600

the visual image is created on the surface. The electrical output of the tube is obtained by scanning the photosensitive surface with an electrical beam as shown in Figure 2-18.

The entire image is captured when the surface is exposed to the illumination flux. The electrical charge image is converted to an analog voltage signal by sweeping the scanning beam across the surface. The charge on the surface modulates the current in the circuit. The beam is stepped from the top to bottom in a standardized pattern to produce the RS-170 output format.

In a system where the camera output is in RS-170 format, the signal can be transmitted directly to a standard TV monitor to produce a visual image on the screen. The video standard (known as RS-170) established by the Electronic Industry Association will be described in detail in a separate section.

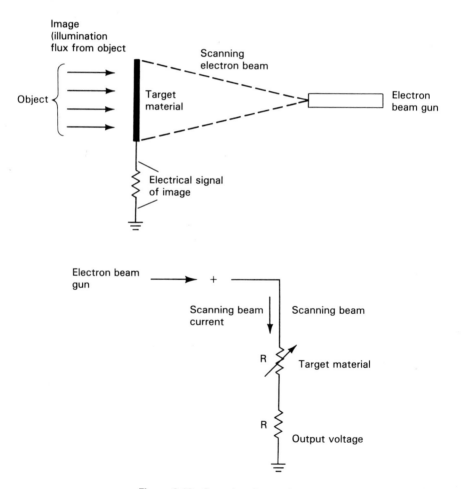

Figure 2-18   Scanning electron beam.

The output of the tube in the camera is an analog signal where the voltage amplitude is proportional to the charge density, and hence to the intensity of the illumination on the area covered by the scan line of the image (Figure 2-19).

The material in the target screen of the tube can be modified to affect the response time of the surface. However, the dynamic response requirement is different for a camera capturing an image of a moving object as contrasted to a camera obtaining an image of a stationery object in low illumination. The moving object requires as short an exposure as possible to reduce blurring, while the latter condition requires as long an exposure as possible to maximize the number of light photons captured. The response time of a high-sensitivity surface will be low with the result that the image will lag and the **edges** in the image will be blurred if the object is moving rapidly.

The screen material and the camera are affected by environmental conditions such as very high intensity illumination (called blooming), by stray electromagnetic radiation, and by temperature.

Tube type cameras may weigh several pounds, due to weight of such components as deflection coils; in contrast, solid state cameras may weigh under a pound. The power consumption of industrial tube type cameras is usually about 10 to 20 watts as contrasted to 1 to 3 watts for the solid state units. Resolution of tube cameras is generally higher than that of solid state devices, but is increasing rapidly in the solid state units as the VLSI capability to produce higher density sensor arrays improves.

Characteristics.   Electron tube cameras are the detector of choice for applications requiring high resolution, color quality or low light intensity sensitivity. The vidicon is the most common type used today, but newer types are gradually replacing it. The following eight characteristics are common to tube type units.

1. High resolution is possible with non-RS-170 format and high sampling rates.

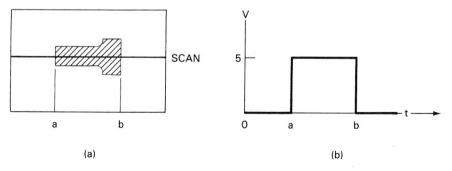

(a)   (b)

Figure 2-19   Analog voltage output signal (b) of tube-type camera scanning image (a).

2. High sensitivity depends on the type of tube.

3. Frequent calibration is required, due to drift and aging.

4. Image lag is produced due to low speed of response.

5. Image burn can occur if there is intense radiation.

6. Units are heavier, larger, and produce more heat than solid state units.

7. Units are affected by magnetic fields and suffer from geometric distortion.

8. Units are sensitive to vibration.

There are fewer new developments in tube type camera technology than solid state units because it is a more mature technology. Getting replacement parts will become more difficult as the solid state units take over more of the applications.

Tube type cameras normally used in the machine vision industry have a C-mount lens.

Analog to Digital Conversion.   Conversion of the **analog signal** from the tube type camera to a digital signal format requires three steps.

1. The analog signal must be sampled.

2. The value of samples must be quantified to the appropriate gray levels.

3. Quantified values must be digitized.

Sampling is the process of obtaining the instantaneous value of the analog voltage at instants of time called sampling points. An image with sharp changes in contrast contains high frequency components and hence will require a higher sampling rate than that for an image not having sharp changes. The higher sampling rate generates more data and, therefore, requires an array with a larger number of elements to obtain the data as well as system bandwidth sufficient to process the data in a timely manner. The speed of the system to enhance image data and the quality of the **image enhancement** will be affected by the sampling rate. The image is enhanced by processing techniques that accentuate certain properties to improve sharpness and decrease interference.

Quantification of the sampled values of voltage is necessary to establish the number of gray levels which will be used by the system. The pixel values must be integers, and rules for rounding must be established. Rounding will be covered in a later section. Note that the rounding process is related to the variance of dimensions based on the data.

Digitizing the quantized data is necessary for it to be used in the vision processing system. Word length capabilities of the system must be equal to or greater than the number of bits required to encode the maximum gray level value.

The resolution of the system is affected by the analog to digital process. The greater the number of samples, the better the resolution of the system, but at the expense of increased bandwidth, computational requirements, and data storage. Analog to digital elements, used to convert the analog signal from the tube type camera to a digital format, are relatively small and may be located either in the camera or in the frame grabber of the vision system.

Sampling.   The rate at which the analog signal must be sampled to obtain the necessary information will depend on the dynamic nature of the image and the number of elements per row in the pixel map. The sampling rate should be at least twice the highest frequency of pixel value changes in the image.

The portion of the continuous analog signal from the tube type camera in RS-170 format shown in Figure 2-20 represents the 52 microsecond period that the electron beam is scanning across the image. The continuous output voltage must be converted to discrete numbers of samples corresponding to the elements of the pixel map of the image. Hence, ten samples must be obtained during the forward sweep time as shown by the arrows in Figure 2-20 for the ten elements in the row of the 10 × 10 matrix of the image.

The sampling time period is

$$\begin{array}{l} \text{Sampling time} \\ \text{(for a row of 10} \times \text{10 matrix)} \end{array} = \frac{\text{Length of active analog signal}}{\text{Number of samples}}$$

$$= \frac{52 \text{ microseconds}}{10 \text{ samples}}$$

$$= 5.2 \text{ microseconds}$$

The first sample could be taken any time during the first 5.2 microseconds of the trace period and the nine subsequent samples taken in uniform intervals of 5.2 microseconds from the time of the first sample

$$\text{Sampling rate} = \frac{\text{Sample}}{\text{Time between samples}}$$

$$= \frac{1}{5.2 \text{ microseconds}}$$

$$= 0.196 \times 10^6 \text{ samples/second.}$$

or can be determined on the basis of the entire trace.

$$= \frac{\text{Number of samples per active trace}}{\text{Time duration of active trace}}$$

$$= \frac{10 \text{ samples}}{52 \text{ microseconds}}$$

$$= 0.196 \times 10^6 \text{ samples/seconds}$$

The data for the first row of a 10 × 10 matrix can be obtained by sampling the first active trace of the image signal at a rate of 0.196 × 10 samples/second and the values transferred to the correct storage locations in

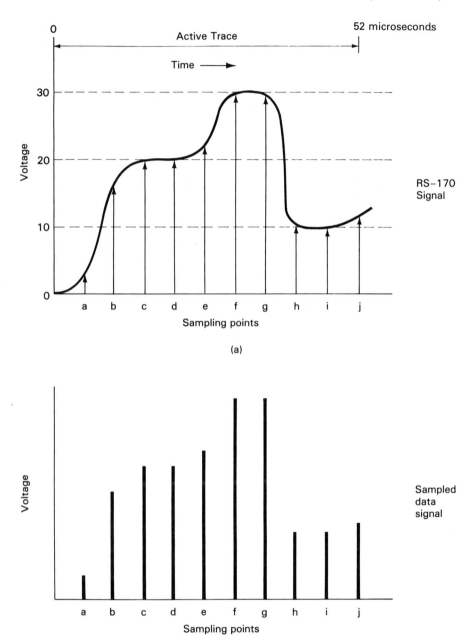

Figure 2-20 Sampled data of analog signal. RS-170 analog signal (a) and corresponding sampled data voltage pulse train signal (b).

the image matrix. It can be seen that only 100 values out of 4,677 values generated by sampling the active part of scans during one thirtieth of a second are used for the 10 × 10 pixel matrix of the image.

The values for the second to tenth rows of the 10 × 10 matrix are obtained by sampling the appropriate trace or averaging an appropriate number of traces. Since there are ten rows in the matrix and a noninterlaced frame takes 0.033 seconds, the second row will be obtained from the trace that starts at 0.033/10 or 0.0033 seconds after the start of the frame. The sweep rate for RS-170 is 15,750 hertz; hence, each total sweep (forward and backward) equals 63.5 microseconds. The trace or scan to provide data for the second row of the matrix is determined by dividing the frame time by ten times the sweep time.

$$\text{Sweep number} = \frac{\text{frame time}}{10 \times \text{sweep time}} = \frac{0.0033}{10 \times 63.5} = 52.5 \text{ sweep}$$

Data would be collected from the 53rd sweep for the second row of the matrix. Data for subsequent rows would be obtained from sweeps indicated in Table 2-2 by following the same process. The sweep number for the third row would be calculated as follows.

$$\text{3rd row Sweep Number} = \frac{2 \times 0.0033}{10 \times 63.5} = 105 \text{ sweep}$$

More data can be used to determine the matrix values by employing a procedure which calculates an average value obtained from a number of sweeps to reduce noise. For example, having sweeps 1, 2, 3, . . . 52 for the first row, 53, 54, 55, . . . 104 for the second row, and so on, would provide an acceptable alternative process.

Solid state cameras.   The basic concept underlying solid state image sensors is that a separate electrical signal is produced for each pixel or area in the sensor. This is accomplished by locating a photoconductor element which has the property that its resistivity decreases proportionally to the amount of light energy falling on the device in each pixel area. The output signal voltage shown in Figure 2-21 is dependent on the resistance of the photoconductor.

The relationship between changes in light energy and output voltage is not linear and depends on the transfer characteristic of the photoconductor. The ratio of increase in electrical current to increase light energy is referred to as the gamma power.

TABLE 2-2   Matrix Input Data

Row of Matrix	1	2	3	4	5	6	7	8	9	10
Sweep Number	1	53	105	158	210	264	316	370	423	472

Light source

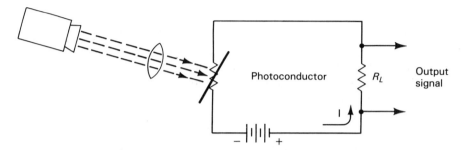

Figure 2-21   Solid state camera photoconductor sensor circuit.

The output of the circuit is a voltage signal corresponding to the value of the light intensity on the pixel being sensed by the semiconductor element. Multiple detectors are used; they are arranged in linear and in rectangular arrays and the image data is obtained by sampling the output of each detector. The detector element must have sufficient sensitivity to detect the incident radiation and a time constant which permits the sampling of all the elements in the array to obtain the image data. The output of the sensor is a series of voltage pulses representing the light intensity at the pixel locations. The voltage pulses must be digitized, if the data is to be entered into a computer, or must be converted to RS-170 format, if the image is to be displayed on a video monitor. Solid state cameras are not subject to blooming and flare problems. A bright light can be directly viewed with minimal probability of chip damage.

Four main types of solid state material used in sensing chips are

1. MOS—Metal Oxide Semiconductor
2. CCD—Charge Coupled Device
3. CID—Charge Injected Device
4. CPU—Charge Priming Device

TABLE 2-3   Solid State Camera Performance

Chip Type	MOS	CCD	CPD/CID
Sensitivity (minimum Lux)	10	3	5
Dynamic Range (relative)	1 (poorest)	4	1
Dark Current	good	poor	good
Noise	15	1	4

CID and **CPD** are very similar and will be grouped together in the performance comparisons (Table 2-3).

The characteristic of an entire row of pixels being inoperative if one pixel fails in **CCD** cameras makes it undesirable for industrial use. In addition, the CCD units are subject to moire pattern which appear as a series of rings on the monitor.

Physical size of an integrated circuit array containing 256 elements in a row would be less than one quarter of an inch in length and contain a ground-polished window for protection of the photodiode element.

Solid state photodiodes can be fabricated in linear or rectangular arrays. Linear arrays can also be fabricated in the form of a circle of equally spaced elements for applications in focusing, tracking, and angle discrimination. A circle with 720 elements would provide 0.5 degree spacing.

Solid state cameras can be in the few ounce weight category and are about one fifteenth the weight of comparable tube type units. Power consumption is in the two watts range and can be powered by a 12 volt battery. The units are relatively shock and vibration resistant and can be mounted in any position. The service life is high with an estimated maintenance-free life of seven to ten years.

Characteristics.   Solid state cameras can contain either line scan or **matrix array sensors.** The latter is composed of multiple rows of light-sensing elements that produce a two-dimensional image.

Characteristics of solid state units are the following.

1. Matrix array cameras are available with RS-170 format output for display on a TV monitor.
2. Line scan units can be attached end to end to produce longer linear arrays; the matrix array size range from three to five hundred by three to five hundred pixels and are increasing with time.
3. Response is highest in the infrared region of the spectrum and may require an infrared correction filter for normal images.
4. Cameras and sensing arrays are available with different levels of bad or malfunctioning pixels depending on the price.
5. There is virtually no geometric distortion, drift, or lag.
6. Cameras are light and rugged and consume little power; they can be operated from batteries.
7. Pixel transfer rates can be as high as 20 megapixels per second.
8. They can have C, U, or bayonet lens mounts with adapter ring for a C-mount lens.

2.2 PROCESSING

Physical limitations of the hardware used to acquire the image introduce both random and systematic noise into the image data. In addition, the image may contain data on features which are not of interest or which mask the items of interest. A primary function of the **image processing** is to create a new image by altering the data in such a way that the features of interest are enhanced and the effects of noise are reduced or eliminated. Specific data processing techniques to accomplish these objectives will be covered in detail in chapters three and four.

The processing elements contain the electronics and software programs to perform

1. Image grabbing
2. Image enhancement
3. Feature extraction
4. Output formatting

The data processing in a specific machine vision system may be performed by either hardware or software. In general, the use of hardware to perform the operation results in a faster system once a desired data processing algorithm has been finalized. The hardware increases the initial cost, reduces the ability to make changes in the processing, and increases the probability that all the similar systems in a manufacturing plant are performing the same processing.

Hardware elements associated with the signal digitizer and frame grabber storage memory are contained in every system. It is necessary to verify that the hardware and any proposed commercially available vision processing software programs are compatible since the operating characteristics have not been standardized.

Basic processing algorithms for common techniques such as noise elimination, edge enhancement, filtering, and gray scale modification are contained in most commercially available vision systems. It is not necessary to write the software to accomplish these basic tasks. Three primary modes of altering the data are

1. Point by point in one image
2. Using corresponding points on different images
3. Using regional points in one image

The point-by-point mode maps each pixel of the original image into a new image where the pixel value of each pixel in the new image is related to the value of the corresponding pixel in the original image. An example would be the inversion of a binary image where the zero pixel values are changed to one and the one values are changed to zero.

The corresponding-points mode creates a new image by the **correlation** of each pixel to the values of corresponding pixels in two or more source images. The values can be combined in different ways as follows. Pixel values of two images of a city taken at different times can be subtracted to determine changes that occurred during the intervening period. Image data can be modified for calibration purposes where variation is due to location; the center may not require any correction, but the outer areas may have to be increased by seven percent. This mode can also be used to combine data from a number of different sensors to form a map of a region based on the composite data. The sensors may be satellite cameras each taking an image in a different portion of the spectrum such as infrared, ultraviolet, and visible.

Regional-point mode involves the calculation of a value for a pixel location in the new image based on the values of a number of pixels in the **region** adjacent to the point. This mode can be used to average data from pixels in the region adjacent to a flawed point to minimize bad data elements. The value of the pixel in the new image would be equal to the average value of the nine pixels in the region adjacent to the corresponding pixel. Most of the image enhancement techniques use the regional-point processing mode.

2.3 OUTPUT OR DISPLAY

Commercially available vision systems will usually have the output capability of data printout, data display, and the generation of control signals. The system provides the user with large volumes of accurate data to assist in making process management decisions. Since the specific data of interest and the method of using the data are dependent on the manufacturing application, the degree of automation at the installation, and on the management structure of the organization, the output will be highly customized.

An on-line video display of the image along with the capability of displaying the **histogram,** a graph showing the frequency at which each intensity occurs is normally available. In addition the system designer can identify special display features for use by the operator, such as, cursors to identify positions on the object, numerical dimensions of features, perimeter values, dimensions of major or minor axis, and classification of the object.

Examples of customized output are the initiation of an output control signal when an object under inspection (a label on a bottle or a part number) does not meet specifications or the periodical printing of monitoring reports for guidance to operators in making adjustments to production equipment.

Vision system output tools available on relatively low cost manufacturing systems will include items such as calipers, key point locators, tape measures, protractors, template tools, arc tools, pixel counters, and defect finders. Each tool is designed to perform a specific task and must be tailored to the application. For example, the purpose of the Vernier caliper is to precisely measure the internal and external dimensions between parallel edges of parts to a precision much finer than the size of the pixels by using subpixel techniques.

The vision system provides a continuous, reliable set of status data on the production process, the ability to statistically upgrade the control values, the capability of displaying the status information in either video or hard copy, and the ability to input manufacturing data to the computer system on a continuous basis.

REFERENCES

1. Kenneth R. Castleman, *Digital Image Processing,* Englewood Cliffs, Prentice Hall, 1987.
2. Edward R. Dougherty and Charles R. Giordina, *Image Processing: Continuous to Discrete,* Englewood Cliffs, Prentice Hall, 1987.
3. Larry Werth, "Automatic Vision Sensing in Electronics Hardware," *Sensors,* December, 1986.
4. H. E. Schroeder, "Practical Illumination Concepts and Techniques for Machine Vision Applications," in *Robotic Sensors,* Vol 1, IFS, Kempston, England, 1986.
5. R. C. Dunn, *Fiber Optic Lighting for Machine Vision,* Society of Manufacturing Engineers, MS-85-226, Dearborn, Michigan, 1985.
6. *CCD: The Solid State Imaging Technology,* Fairchild Western Systems Inc., CCD Imaging Division, 810 West Maude Ave., Sunnyvale, CA 94086, 1988.
7. *Image Sensing Products:* EG&G Reticon, Sunnyvale, California, 1988.

EXERCISES

1. Determine focal length, magnification, depth of field for an industrial installation inspecting flat plates on a moving conveyor belt with front lighting. You have a solid state camera with 100×100 array sensor. Sensor dimensions are 0.30×0.30 centimeters. Distance between the lens and work place is 24 inches. f stop is 8, and there are 16 gray levels and 30 images per second. Object dimensions are $3 \times 3 \times 0.5$ inches. Objects are 18.42 inches apart on the conveyor, and the conveyor is moving 1.25 feet per minute. The object image occupies 50% of sensor array.

2. Determine the number of pixels the object moves in the x, y, and z directions between two successive images from Problem 1.

3. Determine if the speed of the conveyor belt in Problem 1 can be increased to 2.5 and 10.0 feet per minute; explain the reason for your answer.

4. Determine if the system in Problem 1 can be used to inspect objects which measure $6 \times 6 \times 0.75$ inches. If not, how can the set up be changed?

5. Determine if, in Problem 1, a zoom lens with a 1.5 to 3 inch focal length range and the f stop adjustable from 1.8 to 16 can be used for the application.

6. Determine the dimensions of the largest object that can be imaged by a vision system with a magnification of 0.1, a 0.2×0.2 inch sensor array with 50×50 elements, the distance from the object to the lens is 24 inches and the f stop is 16. The object is a plate, cylindrical in shape.

7. A system is set up to inspect the top surface of flat 2×3 inch rectangular objects having vertical dimensions between one quarter and three quarters inch. Objects are on a conveyor belt moving at four feet per minute. Camera has an f stop of 8, a

50×50 array sensor with dimensions of 0.2×0.2 inches, and takes 30 images per second. Determine the focal length of a C-mount lens if the distance between the camera lens and the conveyor belt is (a) ten inches; and (b) three feet.

8. Determine if a zoom lens with focal lengths from 12.5 to 75 mm and f stops from 1.8 to 16 be used for the lens in Problems 1 and 7.

9. Describe how strobe lighting can be used in a back-lighting configuration.

10. Describe how structured lighting can be used in a back-lighting configuration.

11. Sketch the picture on a TV monitor in RS-170 mode produced by a tube camera in noninterlaced mode of a four inch square object; what is the height and width of the image.

12. Draw the output of a tube type and the fifth row of a solid state camera when they view an object on a back-lighted table. The tube camera has RS-170 output and the solid state camera has a 10×10 array; maximum output in both cases is 15 volts for a white area and zero for a black area. Work table surface is ten inches wide and one hundred inches long. Determine the width of the scan beam.

13. Match the columns.

1. Back lighting	_____	a. Lighting technique used to determine shape of object from the intersection of the projected pattern and the object.
2. Front lighting	_____	b. Ratio of the height of an object in the image to the height of the object.
3. Structured lighting	_____	c. Radiation from surface at angle perpendicular to the surface.
4. Specular	_____	d. Camera and illumination source are on the same side of the object.
5. Diffuse	_____	e. Short burst of high intensity light.
6. Tube type camera output	_____	f. Camera and illumination source are on opposite sides of the object.
7. Magnification	_____	g. Radiation from a surface at an angle equal to the angle of the incident radiation.
8. Focal length	_____	h. Continuous voltage analog signal.
9. Strobe light	_____	i. Distance between the center of the lens and the focal point.

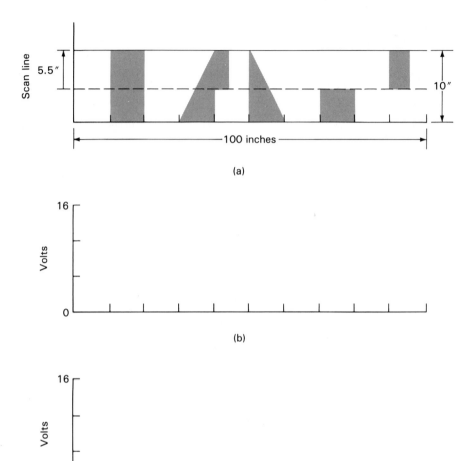

(a)

(b)

(c)

Figure 2-22 Output signals of object on backlighted inspection table (a), produced by a solid state camera (b), and by a tube camera (c).

14. Match the columns.

1. Depth of field	_____	a. Defects appear as bright spots in image.
2. Aperture	_____	b. Structured, strobe, back, and front.
3. Dark field	_____	c. Dimension of sensor array divided by the number of sensors.

4. Phases of image processing   _____   d. Lens opening where the highest "f stop" number represents the opening with the smallest area.

5. Solid state camera output   _____   e. Defects appear as dark spots in image.

6. Phases of image acquisition   _____   f. Image enhancing, feature extracting, and output formatting.

7. Types of lighting   _____   g. Series of voltage pulses where height of each pulse represents the value of illumination on the area represented by the pixel.

8. Light field   _____   h. Region or zone where the object is in focus.

9. Pixel size   _____   i. Illuminating, focusing, and sensing.

3

FUNDAMENTAL
CONCEPTS
OF IMAGE PROCESSING

3.0 INTRODUCTION

The first part of this chapter will present the nomenclature and notation to be used in the second part of the chapter where processing functions will be discussed. The processing of images in digital form shall be considered with one of the following objectives:

1. Enhancing the features or
2. Extracting information.

3.1 PIXEL

The image will be described by an $N \times M$ matrix of pixel values (the elements $p\,(i,j)$ are nonnegative scalars), that indicate the light intensity of the flux on the picture element at (x,y) represented by the pixel. This is illustrated in Figure 3-1, which gives the relationship between the picture element and pixel matrix. The origin in the picture and the matrix are different; the x and y coordinates in the picture start at the lower left corner, whereas the numbering of pixels starts at the upper left corner of the matrix.

$$\text{Where} \qquad i = x \text{ where } 1 \leqslant i \leqslant N^* $$
$$j = (M - y)\; 1 \leqslant j \leqslant M^* $$

* Some PC programs use (0,0) instead of (1,1) for the first location in Figure 3-2. All matrices used in this chapter will be square ($N = M$) to simplify notation. In actual applications, N and M will usually be different.

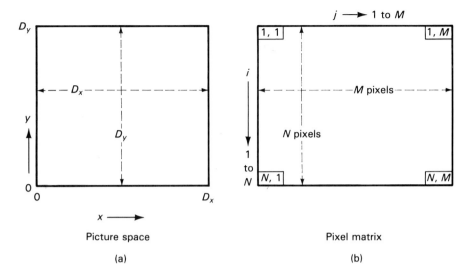

Figure 3-1 Relationship of picture elements (a) and corresponding pixel distribution in matrix (b).

$$x = Dx/N \text{ increment}$$
$$y = Dy/M \text{ increment}$$
$$N = \text{maximum number of pixels in a column}$$
$$M = \text{maximum number of pixels in a row}$$

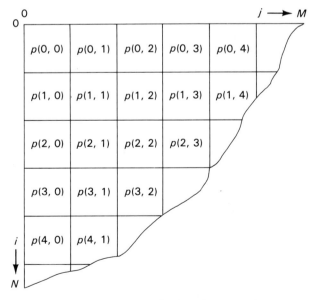

Figure 3-2 Pixel indexing in an image matrix.

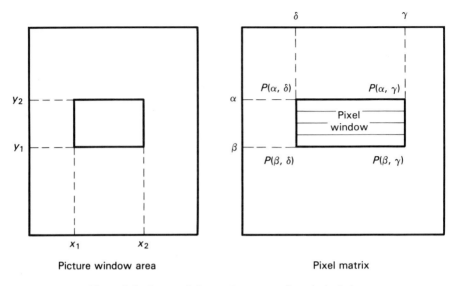

Picture window area   Pixel matrix

Figure 3-3   Image window and corresponding pixel window.

The numerical value or magnitude of the pixel indicates the *average* light intensity on the *picture element area* represented by the pixel. The value of pixel $P_{i,j}$ ranges from 0 to 1.

3.2 WINDOW

A subregion of the picture is called a window and will be designated by the four corners and the corresponding pixel values $P(\beta,\delta)$, $P(\beta,\gamma)$, $P(\alpha,\delta)$, $P(\alpha,\gamma)$ as illustrated by Figure 3-3.

3.3 PIXEL LOCATION

A specific pixel in its most elementary form is identified by its coordinates in the $N \times M$ array representing the picture. The pixel at the (n,m) location has a numerical value which represents the average value of illumination impinging on the area of the picture represented by the pixel.

Consider an example where there is no light or illumination on the top region of a picture and the brightest light available falls on the bottom region of the 10 by 10 inch picture (Figure 3-4). A binary system will be used to represent intensity of illuminations; regions with no light will be indicated by zero and those with brightest illumination by one. The picture will be indicated by a 5×4 matrix, five rows and four columns. Each picture element 2.5 inches wide and 2 inches high will be assigned a number, depending on the average light on the area.

The 2.5 by 2.0 inch area in the upper left hand corner of the picture

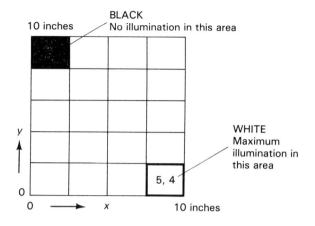

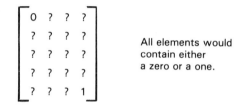

Figure 3-4 Nonuniform lighting on a surface (a) and resultant image pixel values for the surface (b).

represented by location (1,1) in the 5 × 4 matrix would contain a 0, indicating no light. The 2.5 × 2.0 inch area in the lower right hand corner of the picture, that is the fourth column and fifth row, represented by location (5,4), would contain a 1 indicating maximum illumination. Note that the value of the (1,1) pixel would have been 0 and the value of the (5,4) pixel would have been 15 if a 16 gray level system had been used instead of a binary system.

There is no information in the pixel value about intermediate values, and the system designer must indicate a threshold value of illumination where the representation is changed from 0 to 1.

In the example, picture elements are rectangular. The area on the picture could be rectangular or even circular, depending on the sensor. In the case of a tube camera with a circular area sensor, there may be overlapping as shown by Figure 3-5.

In the nonoverlapping case, there are areas of the picture between the circles which are not measured. In case of the overlapping circles, there are areas which are measured twice.

There is no information in the pixel map that indicates the shape of the area in the original picture represented by the pixel or any information about the uniformity of illumination in a given picture element.

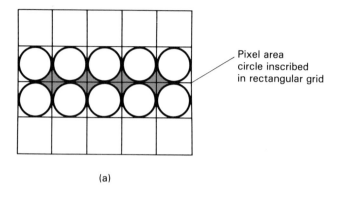

(a)

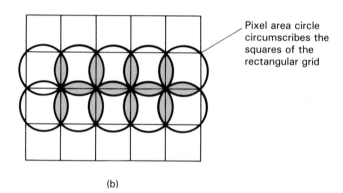

(b)

Figure 3-5 Nonrectangular areas represented by pixel: round nonoverlapping (a) and round overlapping (b).

3.4 GRAY SCALE

In order to provide information on intervening values of illumination, it is necessary to increase the number of bits representing the pixel value. For example, if four levels of illumination are desired, two binary bits are required. Four bits are needed for 16 levels and eight bits for 256 levels. The total number of levels in a gray scale is usually a power of 2. The lowest value 0 is assigned for black, and 1 or a value one less than the maximum gray levels of the system; i.e. 15 for a 16 level system is the value for white. The values assigned to a pixel are always integers.

GRAY SCALE		GRAY VALUE RANGE
2^1	2 values	0, 1
2^3	8 values	0 to 7
2^4	16 values	0 to 15
2^8	256 values	0 to 255

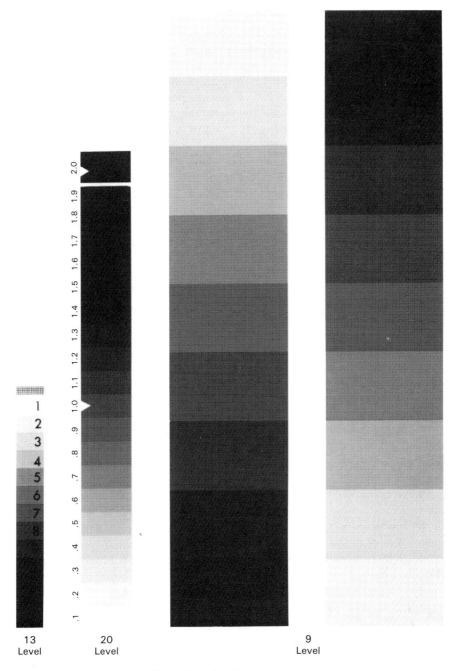

Figure 3-6 Sample gray scales.

The binary representation was used in the early days of vision systems; this simplified the sensors, as well as the data collection, processing, and storage. Today, since most microprocessors are at least eight-bit systems, 16, 64, and 256 levels are common. There is limited value in having more than 256 levels at this time; 64 and 256 levels provide more subtle distinctions than the human eye can see and are sufficient to accommodate current industrial needs. An individual can differentiate between intensities on the order of 40 different levels, but can identify only about 10 to 15. The 16 level gray scale is slightly inferior to the human eye, while the 64 and 256 level gray scale provides for greater discrimination than the human eye.

The system may have the capability of 256 levels, but factors associated with the application may make it necessary to use a specific number of gray levels. The precision or tolerance requirements may best be satisfied by using either a $\frac{1}{9}$, $\frac{1}{13}$, or $\frac{1}{20}$ subpixel resolution technique, and a system with the gray scale range of 9, 13, or 20 would be utilized. A comparison of 9, 13, and 20 gray level scales is shown in Figure 3-6. The number of gray levels affect the image by highlighting certain **features** while eliminating particular details. In general, increasing the number of gray levels improves the quality of the image and provides the opportunity to enhance specific gray regions if the information is in the data base as a multilevel **gray scale image.** The acquisition of images in a binary mode reduces the amount of data storage but limits the ability to use gray scale expansion techniques in the analysis processing operations. Increasing the size of the pixel image from a small value such as 32 × 32 to 250 × 250 increases the resolution and the level of detail in the image. This is different from the zoom function where the pixel is simply increased in size.

The gray scale system provides a means of identifying different levels of illumination intensity in a picture and provide a means of achieving subpixel accuracy in measuring the dimensions of an item. Back lighting is being used to inspect a machine part with a vision system for overall length in Figure 3-7.

The corners of the part in the diagram are in locations (2,1), (2,4), (5,1), and (5,5).

In **cellular analysis,** the area represented by pixel (2,5) is dark region and so would have the value 0 for no illumination intensity. The area represented by pixel (5,2) would be half covered by the object; hence, the average illumination over the area is equal (0 + 15)/2 or 7.5. All other areas are light and have a pixel value equal to 15; the maximum illumination intensity in a 0 to 15 sixteen-level system.

Since the pixel value must be an integer, the 7.5 value must be modified. It is necessary to have a rule in any system which determines how to convert a fractional value to an integer. For example, a general rule could be that the value is rounded up to the next higher integer if the fractional value is between 0.5 and 1.0. Hence, if the measured value is 7.5, the value of 8 would be used as in the lower row of Figure 3-7. If the measured value is 6.6, the value 7

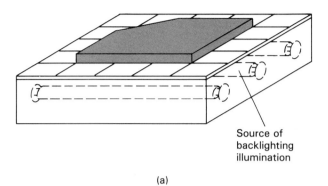

Source of
backlighting
illumination

(a)

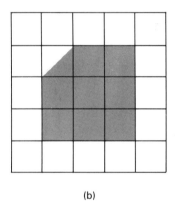

(b)

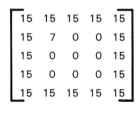

(c)

Figure 3-7 Object on light table (a), with top view (b), and corresponding image matrix (c) where black pixel is represented by zero and white pixel is represented by fifteen.

would be used. The point at which a fractional value is rounded up is of consequence, as it affects the tolerance. It is important to know what that value is for the system being used.

3.5 QUANTIZING ERROR

The rounding of the gray scale values to integer values will result in an uncertainty about the precision of the value. The uncertainty will be a function of the rounding rule.

The rule that any value greater than the integer will be rounded up to the next integer is illustrated by Figure 3-8 for a 16 gray level system. This rounding results in a negative uncertainty.

The dimension and tolerance can be determined by using a subpixel resolution technique, and knowledge of the rounding rule in the following manner. The x dimension in Figure 3-9 is represented by four whole pixels and one partial pixel; the y dimension is represented by one whole pixel.

$$x = 4 \text{ pixel} \times 0.100 \text{ inch/pixel (on object)} + 1 \text{ partial}$$

$$\text{pixel} \times 12/16 \times 0.100 \text{ inch/pixel}.$$

12/16 factor is = proportion of gray levels in partial pixel/total gray levels.

$$x = 0.400 \text{ inch} + 0.075 \text{ inch}$$
$$x = 0.475 \text{ inch}$$
$$y = 1 \text{ pixel} \times 0.100 \text{ inch/pixel} = 0.100 \text{ inch}$$

The tolerance is determined from the rounding rule. A pixel value of 0 represents a full value of illumination, a pixel value of 11 represents a value of illumination from 10.x to 11.0 where x is a value greater than 0.0 and less than or equal to 1. A pixel value of 15 represents a value of illumination from 14.x to 15.0. In summary:

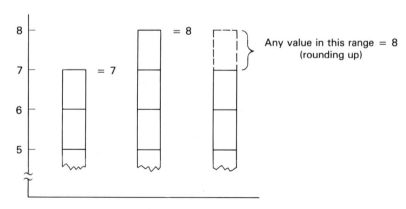

Figure 3-8   Quantizing error (rounding up).

Pixel value in image matrix		Value of illumination represented by pixel value
0	=	0
1	=	$0 <$ pixel value $\leqslant 1$
11	=	$10 <$ pixel value $\leqslant 11$
15	=	$14 <$ pixel value $\leqslant 15$

Hence, the measurement error in a pixel with the rounding rule is such that the object cannot be greater than that indicated by the dimension calculation, but can be less by an amount of pixel dimension represented by one gray level. In the x direction, the two end pixels contribute to the measurement uncertainty, where as in the y direction only one pixel contributes to the uncertainty.

Tolerance or uncertainty of the dimension:
$$x = + 0.0 \text{ inch} - 2 \text{ pixel} \times 1/16 \times 0.100 \text{ inch/pixel} = -0.0125 \text{ inch}$$
$$y = + 0.0 \text{ inch} - 1 \text{ pixel} \times 1/16 \times 0.100 \text{ inch/pixel} = -0.00625 \text{ inch}$$

The dimensions of the object are:
$$x = 0.475 + 0.0 \text{ inch} - 0.0125$$
$$y = 0.100 + 0.0 \text{ inch} - 0.00625$$

The value at which the pixel value is increased to the next higher integer in the round up mode and decreased to the next lower integer in the round down mode is know as the threshold. A threshold in the interval between two integers will result in both positive and negative tolerance.

The rule that values greater than 6.5 will be rounded up to 7 and values between 0.6 and 6.5 (inclusive will be rounded down to 6.0 would result in positive and negative uncertainty zones each equal to

$$\frac{1}{2 \times (\text{max no. gray levels})}$$

The higher the number of gray levels, the smaller the quantizing error.

3.5.1 Measurement Error

The dimension of an object in an image can be determined by counting the pixels or part of the pixels in the direction of interest. If back lighting is used, the image is either black or white, and any intermediate gray scale values are due to the covering of a portion of the pixel area by the object. The precision of the dimension measurement is affected by the quantizing error associated with the value associated with the pixel area at the end of the object.

The dimension of an oblong rectangular object in an image from a backlighted 16 gray-level system with a pixel dimension of 0.10×0.10 inch is

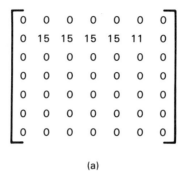

(a)

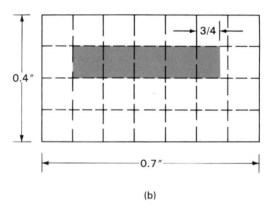

(b)

Figure 3-9 Dimensional determination from image matrix data (a) of regular shaped object on backlighted inspection table (b).

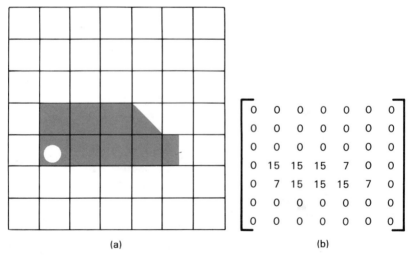

(a)   (b)

Figure 3-10 Irregular shaped object on light table (a) with corresponding image pixel vales (b).

determined by adding the dimensions of the pixels having the maximum gray scale value plus the partial dimension represented by intermediate gray scale values, as shown in Figure 3-9, using the round up to next integer value rule.

There are many different shapes of objects which will result in intermediate gray level values in the back lighting system, as shown in Figure 3-10.

The three pixel areas with the values of seven represent the effect of different part shape characteristics. Area (2,4) is 50% covered by the end of the part at a 45 degree angle; area (3,5) is 50% covered by a rectangular extension of the object, and area (1,3) is 50% covered by the object surrounding the hole. It is necessary to use the known features of the object in analyzing the vision image matrix values for dimensional and tolerance analysis.

3.6 HISTOGRAMS

A histogram is a graphical presentation of the frequency count of the occurrence of each intensity (gray level) in an image. The abscissa or x-axis is the values of gray levels and the ordinate or y-axis is the number of pixels having that gray level (Figure 3-11). The histogram is constructed by

1. Digitizing the image frame,
2. Counting the pixels at each gray scale level, and
3. Plotting the frequency count of pixels at each gray level.

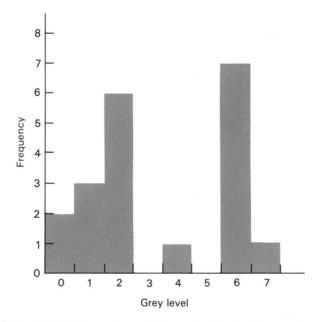

Figure 3-11   Histogram for an eight gray level four by five array image.

The graph can be considered to be a collection of bar graphs showing the number of pixels at each gray level occurring in the frame.

The magnitude of the histogram at a specific pixel value is the probability of the gray value occurring in any picture element in the frame. No information is given about the location of the pixel. The probability of a pixel value b occurring at a given point (x,y) in the picture is given by Figure 3-11 and the equation

$$p(b) \; \frac{\text{at any point } (x,y)}{\text{in the picture}} = \frac{\text{value } b}{\text{total no. of pixels}}$$

If $b = 6$, the gray level and value of the histogram at $6 = 7$ is

$$p(6) = \frac{7}{20} = 0.35$$

The shape of the histogram provides information on the characteristics of the picture. For example, a narrow histogram indicates a lack of **contrast** in the image, or a specific pixel value may represent a unique characteristic of the item in the picture such as a hole in a part. The histogram is useful in setting the threshold level, in converting a gray scale image to a binary image, or in modifying a portion of the gray scale spectrum.

(a)   (b)

Step 1: Determine gray scale values of object.   Step 2: Determine pixel map of image.

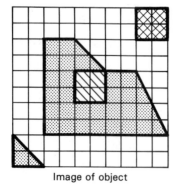

$$\begin{bmatrix} 15 & 15 & 15 & 15 & 12 \\ 15 & 0 & 7 & 15 & 15 \\ 15 & 0 & 7 & 0 & 12 \\ 15 & 0 & 0 & 0 & 3 \\ 7 & 15 & 15 & 15 & 15 \end{bmatrix}$$

Image of object   Pixel image matrix
                                            (5 × 5)

Gray Scale Key
- ☐ White = 15
- ▨ Gray = 12
- ◿ Gray = 7
- ▦ Black = 0

Figure 3-12 Histogram construction: the pixel values of the object (a) determined in step one, and the pixel image matrix (b) determined in step two. A five by five matrix is utilized to simplify the illustrative calculations.

3.6.1 Construction

The histogram of an image can be constructed by the following procedure as illustrated in Figures 3-12 and 3-13.

1. Determine the total number of pixels which will be used to describe the image. This is determined by the array matrix. This example uses an $M \times N$ matrix and the total number of pixels is the product $M \times N$. If

$$M = 10$$
$$N = 10$$
$$\text{Total pixels} = 10 \times 10 = 100$$

It should be noted that M does not have to equal N; this will depend on the type of camera, the analog-to-digital sampling rate, and the memory of the system. The higher the value of M and N, the greater the precision or accuracy but at the cost of increased computational capacity and time of response.

2. Construct a pixel map of the image. In the example, a 5×5 matrix was constructed; the total number of pixels has been reduced to 25.

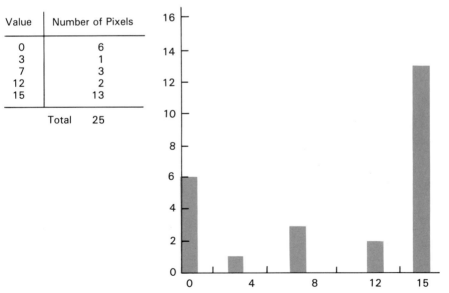

Step 3: Tabulation of pixel values

Value	Number of Pixels
0	6
3	1
7	3
12	2
15	13
Total	25

Step 4: Graph tabulated data

Figure 3-13 Histogram construction: pixel values tabulation (a) of image data in Figure 3-12(b) and the graphical representations of the tabulated data (b).

3. Tabulate the results of the pixel map: that is, count the frequency of occurrence of each pixel value. For example, the value 0 occurs in six pixels, but the value 3 occurs only once. The total number of values in the table must be equal to $N \times M$; in the example, this is equal to 25.

4. Construct the histogram by plotting a bar graph for each pixel value from 0 to the gray scale minus 1. In the example, the gray scale of 16 was used to construct the pixel map, and so the highest value is 15. It should be noted that the maximum value of the pixels in the map is not a factor. The histogram will consist of bar graphs for pixel values 0 to 15.

3.6.2 Characteristics

1. Spac 1AL information is discarded. The histogram provides information on the frequency of occurrence of pixels having the various gray levels but no information on the distribution or location of the pixels in the image.

2. Changing the orientation of the object has no effect on the histogram. In essence the histogram is a measure of the integrated gray scale values over the entire frame

$$\left(\sum_{i=0}^{15} n_i \, gi \right)$$

where n_i is the number of pixels at each gray scale level i, and i is the number or value of the gray scale level. If the value changes when the orientation is altered, it can be concluded that the sampling was not appropriate. The sampling rate should be increased until the histogram is constant.

3. A histogram for a specific image is unique. The image for a specific histogram is not unique.

4. A histogram provides information on the scaling of images in the available range of grays and an insight into rescaling an image. The data can be rescaled to achieve higher contrast of the features of the image.

5. Histograms provide information on equalizing two images for subtraction purposes.

3.7 RGB AND CMYB COLOR SYSTEMS

The color systems will have three or four values associated with each pixel location. For example, in the conventional red, green, blue (RGB) system, there will be three values for each of the three components at each pixel location and the combination of the three on the chromaticity diagram in Figure 3-14 provides the exact specification of the color of the pixel. The printing industry usually uses the four color CMYB system, since the observer is looking at reflected light which is a subtractive process.

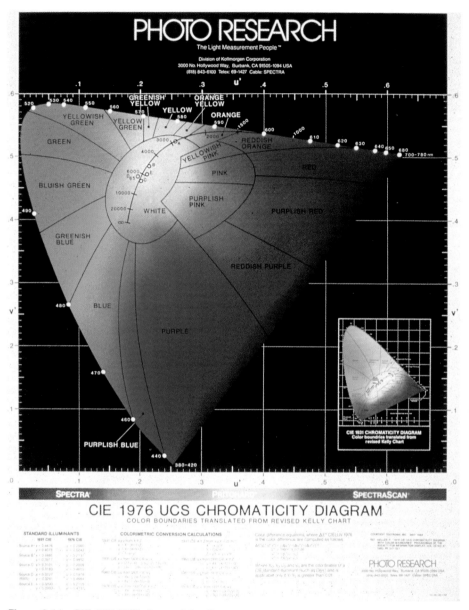

Figure 3-14 CIE 1976 UCS chromaticity diagram.

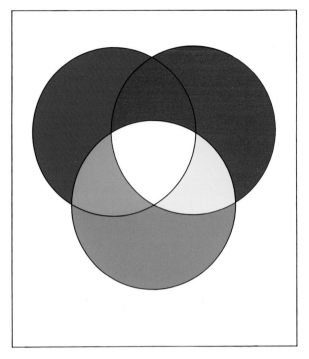

Figure 3-15 Additive colors (RGB). The primary colors, red, green, and blue, when properly additively combined will form white as indicated in the center of the diagram.

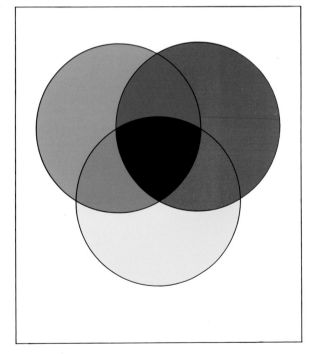

Figure 3-16 Subtractive colors. The proper combination of the secondary magenta, yellow, and cyan colors will produce black as indicated in the diagram.

It is also possible to use the CIE chromaticity diagram in the opposite direction: That is, select a point on the chart which is the desired color; the coordinates of the point will indicate the values to be used. Each of the color components will have a 16 or 256 level value.

The red, green, blue color system used in industry and the four color, cyan, magenta, yellow, and black (CMYB), system used in the printing industry differ in that the former can be considered as additive and the latter as subtractive.

If you observed, as in Figure 3-15, the results of projecting the RGB light beam on a white opaque background projection screen in a dark room, you would note that it is black where there are no beams and white where all three beams overlap. The white area is the result of the presence of all color components or an additive process.

Consider the CMYB system illustrated in Figure 3-16. The white light, containing all components, is projected on a screen. Filters which subtract the given components are inserted between the light and transparent screen. Thus, this is a subtractive process.

REFERENCES

1. Andrew C. Stougoard, Jr., *Robotics and AI*, Englewood Cliffs: Prentice Hall, 1987.
2. Brian Case, "Pipeline Processor Pushes Performance," *Electronic Systems Design Magazine*, March, 1987.
3. Stanley N. Lapidus, "Gaging with Machine Vision," *Vision Technology*, May, 1986.
4. Andrew W. Davis and Imran Khan, "Pyramid Processing Redefines Machine Vision," *Digital Design*, May, 1986.
5. Rita R. Schreiber, "Quality Control with Vision," *Vision Technology*, October, 1985.
6. Gregory A. Baxes, *Digital Image Processing*, Englewood Cliffs: Prentice Hall, 1984.

EXERCISES

1. The inspection table is being used to check a machine part with 16 gray levels. The key for the gray levels is shown in Figure 3-17.

2. Can you determine the type of camera used to produce the image in Figure 3-17? Explain your answer.

3. Assume the dimensions of the inspection table are 2 × 2 inches and that you are in a round up binary system mode with a threshold of 0.9. What is the accuracy of your measurement of the largest vertical dimension? Explain your answer.

4. Using the dimensions in Problem 3, what is the vertical dimension of the part?

5. Using the dimensions in Problem 3, what is the accuracy of the measurement of the overall width of the part? Explain your answer.

6. Construct a pixel matrix for the machine part image in Problem 3 using the 9 level gray scale in Figure 3-6. Identify the transition points used.

7. Determine the gray pixel value in a back lighted binary, 16 gray level, and 256 gray level systems. Round upward. Use the image in Figure 3-17 for a 10×10 matrix.

8. Give the image matrix and determine the dimensions and tolerance of the following objects on a 4×4 inch light table for the binary 16 level and 256 level system (Figure 3-18).

9. Determine the gray pixel value in a back lighted binary, 16, and 256 gray level system for a through d of Figure 3-18. Round upward.

10. Give the image matrix and determine the dimensions and tolerance of the following objects on a 14×14 inch light table for the binary, 16 light, and 256 level system in e and f of Figure 3-19.

11. Determine the x and y dimensions and the tolerance of the rectangular object on the 10×10 inch inspection table with back lighting. Assume black is 15 and white is 0. The 16 level system has a 5×5 sensor array and uses a round up rule. Indicate the effect of using a 256 level system and explain any differences.

$$\begin{bmatrix} 0 & 0 & 0 & 0 & 0 \\ 0 & 0 & 0 & 0 & 0 \\ 0 & 3 & 15 & 15 & 0 \\ 0 & 0 & 0 & 0 & 0 \\ 0 & 0 & 0 & 0 & 0 \end{bmatrix}$$

Input matrix

	16 Level System	256 Level System
x Dimension	_____	_____
Tolerance	+ _____	+ _____
	− _____	− _____
y Dimension	_____	_____
Tolerance	+ _____	+ _____
	− _____	− _____

Determine the effect of using a round-down instead of a round-up rule.

	16 Level System	256 Level System
x Dimension	_____	_____
Tolerance	+ _____	+ _____
	− _____	− _____
y Dimension	_____	_____
Tolerance	+ _____	+ _____
	− _____	− _____

12. Match the columns.

    1. Window   _____   a. Colors which, when combined together in an appropriate manner, result in black.

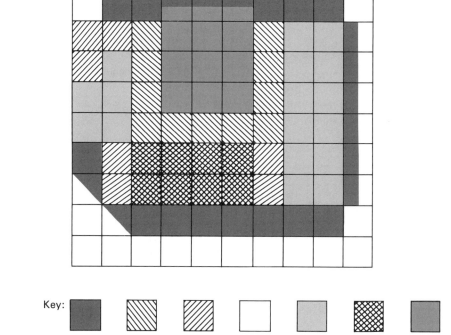

Figure 3-17 Image of machine part on back lighted inspection table.

2. Binary system	_____	b. Colors which, when combined together in an appropriate manner, result in white.
3. 16 level gray system	_____	c. Graph indicating frequency of the occurrence of each gray level in an image.
4. Round up rule	_____	d. Uncertainty of parameter value due to pixel values being limited to integers.
5. Histogram	_____	e. Provides black and white image.
6. Additive colors	_____	f. Procedure of changing noninteger gray level values to the next higher integer value.

7. Quantifying error   _____   g. Sub-region of an image
                                                    designated by the four
                                                    corners.

8. Narrow histogram   _____   h. System with gray scale
                                                    range of values from
                                                    0 to 15.

9. Subtractive colors   _____   i. Image with a lack of or
                                                    low contrast.

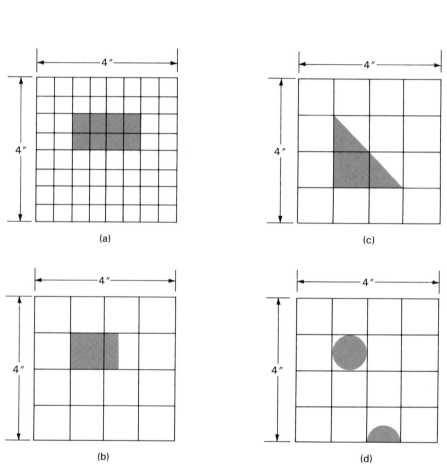

(a)

(c)

(b)

(d)

Figure 3-18 Objects on light table: rectangular shape covering full pixel (a),
rectangular shape with partial coverage of 4 pixel (b), triangular shape (c), and
circular shape (d).

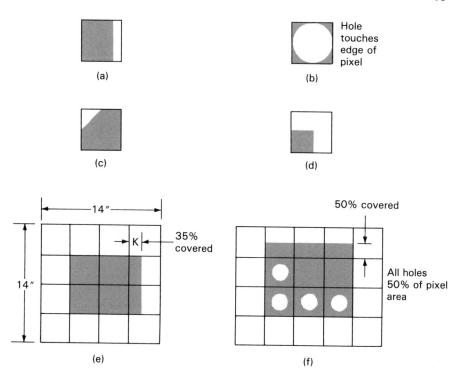

Figure 3-19 Subpixel resolution: object partially covering pixel (a), object having a hole with a diameter equal to the size of pixel (b), object with corner missing (c), square object covering one quarter the area of the pixel (d), rectangular object partially covering 35 percent of the adjacent pixel on one end of the object (e), and complex object partially covering pixels on one edge and having four holes with an area equal to one half of a pixel area (f).

4

IMAGE AQUISITION SIGNAL PARAMETERS

4.1 IMAGE ACQUISITION

The electrical signal representing the image is obtained in two ways, by sequentially scanning with one sensor in tube type cameras and by reading the output of multiple detectors of the array sensor in solid state cameras. A voltage versus time signal is obtained in both cases, but the shape of the two signals is different. The signal contains information about the average flux on each pixel in the image as shown in Figure 4-1. It does not provide any information on the distribution of flux over the pixel area.

The analog video signal in RS-170 format for display purposes on the standard TV monitor used in the United States is shown in Figure 4-2(a). The synchronization signal is contained in the format, hence the signal is fed directly to the monitor.

The binary signal for use in the image processing system is shown in Figure 4-3. A frame-grabber unit acquires the camera data over a finite period of time, organizes it, and stores the data for an entire image in a temporary local storage memory. The image data is then transferred as a unit to the main vision system storage internally or through a standard **RS-232** link.

4.1.1 Synchronization

The camera and frame grabber must be synchronized. This can be accomplished in either of two ways:

1. Sync Lock
2. Sync Gen

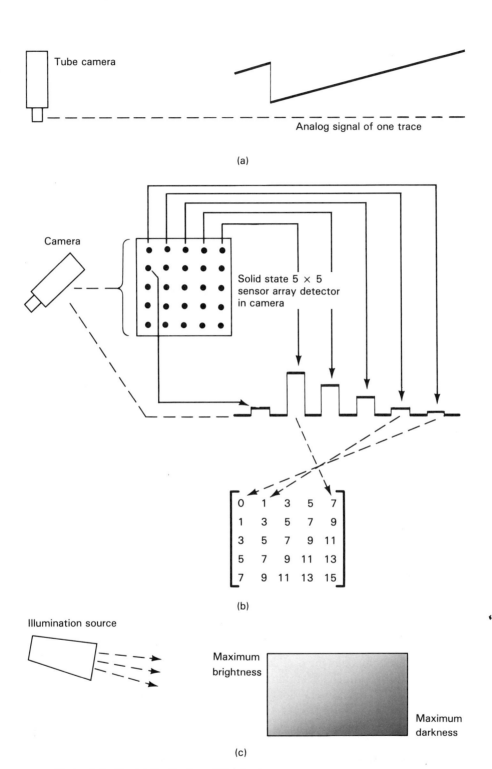

Figure 4-1 Spatial distribution of illumination: analog voltage signal from tube type camera (a), voltage pulse train from (five by five array) solid state camera and resultant pixel values in image array (b), and image with brightness gradient (c).

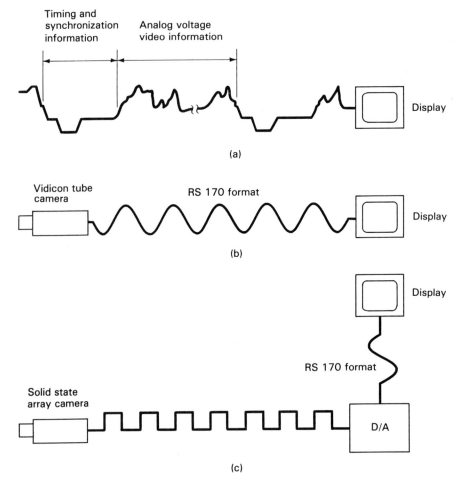

Figure 4-2   Signal conversion for display on standard TV monitor; signal in RS 170 format (a), signal from Vidicon tube camera (b), and signal from solid state array camera (c).

The Sync Lock system uses a phase control loop and requires separate driving signals for vertical and horizontal blanking and other functions. The Sync Gen system locks the oscillator of the camera to the system signal and is the preferred appɪoach as it minimizes the complexity of the system controls (Figure 4-4). The camera specifications generally indicate the type of synchronization required.

The Sync Lock system provides better coordination of various units when this is necessary for adaptive control or other purposes (Figure 4-5). Interference, noise, signal attenuation, and modification of the system configuration are a more serious problem in this system than in the Sync Gen system.

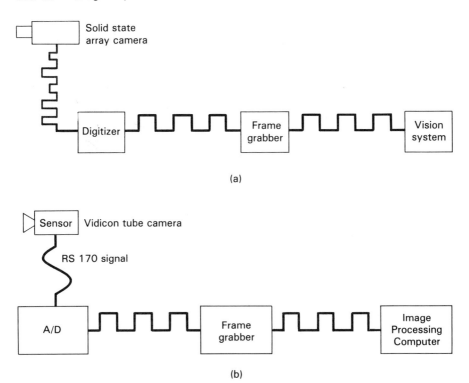

Figure 4-3   Signal conversion for image processor use: solid state camera system (a), and Vidicon tube camera system (b).

The local oscillator of any unit in the system can provide the system synchronization signal. If the composite output of the camera is used for this purpose, it is necessary to strip the synchronization signal from the composite signal for control purposes.

4.1.2 Interlacing

The early industrial vision systems used commercial United States broadcast television equipment to the greatest extent possible to minimize costs. This resulted in the adoption of the RS-170 format developed by the broadcast industry to provide a publicly acceptable picture on home TV sets while using as narrow a bandwidth as possible in order to minimize the frequency spectrum.

The process of generating an image by use of the alternate line field structure required by the RS-170 standard is known as interlacing. The RS-170 format standard, administered by the Electronics Industries Association (EIA), prescribes that the image on the monitor have 480 lines (generated at 525 lines per $\frac{1}{30}$ second rate) divided into two fields of 240 lines. One field contains the odd numbered lines, and the other field consists of the even

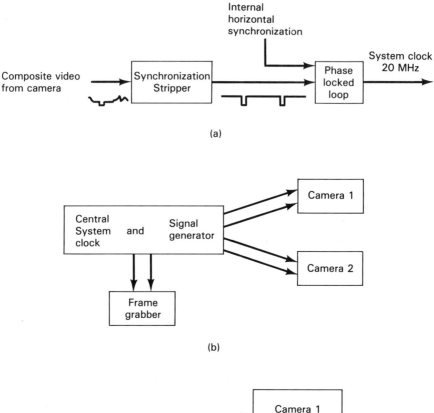

Figure 4-4 Synchronization systems: camera composite video signal used for system clock signal (a), SYNC LOCK (no local oscillators used, central system clock generates specific individual timing control signals for each element) (b), SYNC GEN (local oscillators in each unit used with one central control signal to synchronize system elements) (c).

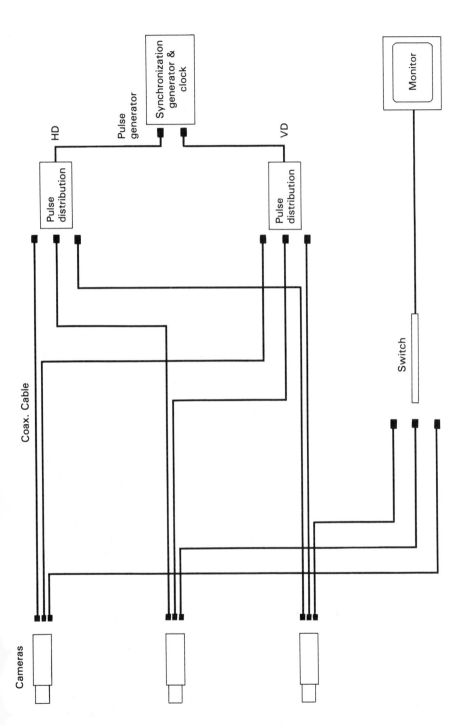

Figure 4-5 SYNC LOCK synchronization of multi-camera system with separate horizontal and vertical drive control signals.

numbered lines. Each field is scanned in 1/60th of a second to obtain a complete image every 1/30th of a second. The overlap improves the resolution of the image but does not have any real impact from the industrial application standpoint.

The need for the equipment to operate in an industrial environment and the growth of the industry in recent years have resulted in the development of special noninterlaced cameras and monitor equipment. This simplifies analog to digital processing and data storage tasks but results in a camera which cannot be used with the standard low cost commercially available TV monitor. Some cameras will operate in either mode.

4.1.3 RS-170 Format

The composite signal specified by the RS-170 format standard contains all the necessary timing, control and analog image data required by television monitors in the broadcast industry to produce a picture. Figure 4-6 illustrates the general characteristics of the signal. The frequency of the scanning is 525 horizontal cycles per $\frac{1}{30}$ second.

The time period for a complete scan cycle consisting of an active forward scan and an inactive return trace is 63.5 microseconds. The duration of the forward active scan is 52.1 microseconds and the return trace is 11.4 microseconds.

The width of a scan line in the image of a 24.0 inch object is determined by dividing the object diameter by the 480 lines in one frame of the image. Hence, each scan represents a 0.05 inch wide strip of the object. Vertical **resolution** is limited by the number of scan lines. Horizontal resolution is determined by the sampling rate. For example, if the data is being stored in a 300 by 480 matrix, it is necessary to sample the active traces 300 times per scan, and each trace corresponds to a row of the matrix. The proposed high definition TV (HDTV) would use 1125 lines per $\frac{1}{30}$ second.

Other broadcast standards which have been developed are NTSC, for color images in the United States, which augments the RS-170 by utilizing a 3.58 MHz color subcarrier on the video signal, RS-343-A which covers images with 675, 729, 875, 945, and 1023 vertical lines, and CCIR which covers the 625 lines at a 50-Hz frame rate of the European monochrome broadcasting system. PAL is the equivalent of NTSC for color images in the European system. The lines per field controls the maximum vertical resolution whereas the horizontal resolution is determined by the sampling frequency.

4.2 IMAGE INFORMATION

Edges are normally key information in human perception of images. A square is a closed figure comprised of four edges of equal length arranged in a form having four rectangular corners, while a triangle is a figure consisting of three edges connected by nonrectangular corners.

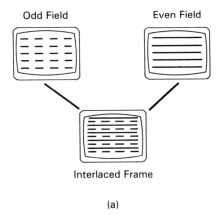

(a)

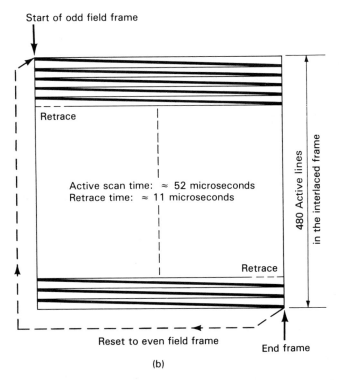

(b)

Figure 4-6   RS-170 video signal format: interlaced display (a), and TV picture frame parameters (b).

The measurement of the length of an object in an image is accomplished by determining the location of two edges denoting the ends of the object, and counting the pixels between them. The measurement of the area of an object is simply an expansion of the linear measurement process and is accomplished by counting the pixels bounded by the four edges. The problem is that edge information may be masked by noise in the image. Hence, an important task of the vision system is to process the image data in a manner that will enhance the edges so that they can be identified for use in feature extraction.

The movement of an object within an image is detected by subtracting one image containing the object from an identical image taken at a different time. The resultant image will have edges or lines where there has been a change. The number of pixels between the lines resulting from the subtraction process is a measure of the movement.

Some parameters based on standard mathematical techniques which can be used for the characterization of an object are listed here.

FEATURES USED TO RECOGNIZE THE OBJECT

Perimeter: Number of pixels in the perimeter of the total object.

Total pixels: Number of pixels contained within a given perimeter.

Bounding rectangle: Outside limits of the area or region.

Circularity: Measure of roundness of the object or region.

Total holes: Count of the number of holes or special regions.

Center of gravity: Location of the weighted average or center of area.

First moment: Moments with respect to given axis; rotationally dependent. A moment is the pixel value × distance from a point on the axis.

Second moment: Square of first moment divided by the total number of pixels in the area; rotationally dependent.

First moment (COG): Moments around the center of gravity; rotationally independent.

Length and width: Diagonal of the object in specific directions.

Histogram: Shape and distribution of intensity values.

The relevance of each parameter for an application such as object identification must be determined by the system designer. The variance determined from a number of measurements on similar objects will provide insight on the value of the parameter for the identification task. The smaller the variance, the greater the degree of certainty of the identification. Since the information on the parameters identified in the preceding list is related to edges, a main objective in processing image data is to amplify changes between adjacent pixel values in the region of the image containing the edge. a noise reduction technique can frequently improve edge detection.

The noise can be in the form of a bad pixel data point due to the failure of an element in the sensor array. The bad data at a given location can be replaced with an estimated value based on averaging the values of adjacent pixel using the dyadic techniques covered in Chapter 6. The averaging technique illustrated in the following example is normally very effective because the pixel values do not change drastically between adjacent pixels. At the same time, it must be realized that the averaging process will tend to broaden the width of edges in the image. A set of conventional techniques for feature enhancement will be presented in detail later.

$$\begin{bmatrix} 9 & 10 & 10 & 11 & 12 \\ 9 & 10 & 10 & 12 & 13 \\ 9 & \text{⑦} & 10 & 13 & 14 \\ 9 & 10 & 10 & 13 & 13 \\ 10 & 11 & 11 & 14 & 12 \end{bmatrix}$$

Bad Data Element in Matrix

Correct the bad data element in the matrix by letting ? be the average of four surrounding elements rounded up to the next integer value.

$$? = \frac{9 + 10 + 10 + 10}{4} = 9.75 = 10$$

$$\begin{bmatrix} 9 & 10 & 10 & 11 & 12 \\ 9 & 10 & 10 & 12 & 13 \\ 9 & \text{⑩} & 10 & 13 & 14 \\ 9 & 10 & 10 & 13 & 13 \\ 10 & 11 & 11 & 14 & 12 \end{bmatrix}$$

Corrected Data Element in Matrix

REFERENCES

1. William Green, *Digital Image Processing,* New York: Van Nostrand Reinhold, 1982.
2. L. R. Rabiner and R. W. Schafer, *Digital Processing of Speech Signals,* Englewood Cliffs: Prentice Hall, 1978.
3. *EIA RS-170 Standard,* Electronics Industries Association, 2001 I Street NW, Washington, DC 20006.
4. Larry Werth, "Automated Vision Sensing in Electronic Hardware," *Sensors,* December 1986.

EXERCISES

1. Sketch the image of a square object which will appear on a monitor if the camera is operating in noninterlaced mode.
2. Draw a flow diagram for the display of an image on a monitor for a system with a solid state camera.

3. Draw a control system for a one camera system. Discuss reasoning for your solution.

4. Draw a control diagram for a three camera system. Discuss the reasoning for your solution.

5. What parameters would you recommend for an inspection system of washers where the size of the center hold is critical? Discuss the reasoning for your recommendation.

6. What parameters would you recommend for an inspection system of washers where both the outside size and the center hole size are critical? Discuss reasoning for your solution.

7. Match the columns.

1. SYNC LOCK	_____	a. Uses local oscillators in units.
2. SYNC GEN	_____	b. Electronic Industries Association (EIA) standard for TV signals, interlaced, 1.4 volt peak to peak, at rate of 525 scans per 1/30 second.
3. RS-170 Format	_____	c. 52 microseconds.
4. Frame grabber	_____	d. 63 microseconds.
5. Number of RS-170 scans in TV image	_____	e. Perimeter, total pixels, bounding rectangle, center of gravity, length, width, ...
6. Number of scans per 1/30 second	_____	f. 480.
7. Number of fields in RS-170 frame	_____	g. 525.
8. Product identification parameters	_____	h. Unit with local memory which stores entire frame of image data from camera before passing it on to the processing system.
9. Duration of active portion of RS-170 scan cycle	_____	i. Two interlaced fields.
10. Duration of full RS-170 scan cycle.	_____	j. Uses central system oscillator.

5

BASIC MACHINE
VISION PROCESSING

5.0 INTRODUCTION

The processing of data in the vision system can be categorized into

1. Point by point (monadic) alteration of data on a global scale, and
2. Multiple point (dyadic) determination of elements of a new array of the image.

   The generation of the new pixel image matrix will be a function of either individual pixel location values or values of pixels in the neighborhood of the individual cell, as indicated in Figure 5-1.

   Monadic operations involve the generation of a new array by modifying the pixel value at a single location based on a global rule applied to every location in the original array. The process involves obtaining the pixel value of a given location in the array, modifying it by a simple linear or nonlinear operation, and placing the new pixel value in the specific corresponding location of a new array. The process is repeated on the pixel value of the next location and continued over the entire array (Figure 5-2). As illustrated by Figure 5-2, the monadic operator is a one-to-one transformation. The operator f is applied to every pixel in the image or section of the image, and the output is dependent only on the magnitude of the corresponding input pixel; the output is independent of the adjacent pixels. The function transforms the gray level value of every pixel in the frame, and the new value is defined by the equation

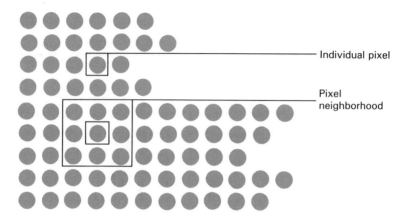

Figure 5-1 Point and neighborhood functions.

$$g(i,j) = f[p(i,j)]$$

The function f can be either a linear or nonlinear operator. The mathematical process is relatively simple in that it involves obtaining the product of two values. The resultant image matrix will be the same size as the original image matrix.

Dyadic operations use essentially the same procedure except that a new image array is generated in one of two ways.

1. The value of each pixel in the array is dependent on a combination of the values of the corresponding pixel locations in two or more similar image frames.
2. The pixel values in the region adjacent to the location of the pixel value being generated are combined.

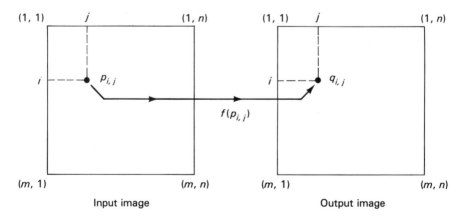

Figure 5-2 Monadic operator.

The latter procedure is accomplished by the convolution process. Convolution can involve any size matrix, but a 3 × 3 matrix is the most common as it requires less computational time and computer capability than a larger matrix. In its simplest form, the convolution of two matrices, p and q, involves the multiplication of all the corresponding elements. The sum of the products of the elements is multiplied by a normalization constant to keep the resultant pixel value within the desired range. The resultant pixel value is inserted in the appropriate location of the new array, and the procedure is repeated after the coordinates of each element are shifted by one in either the M or N direction until all the locations in the array are covered.

5.1 MONADIC ONE POINT TRANSFORMATION

Monadic point-by-point operators are the simplest or most elementary image processing operation. The process involves a single input image (p), as shown in Figure 5-2, and results in a single image (q) output. The point-to-point operators which will be covered in this section are the identity operator, inverse operator, threshold operators, and combinations thereof.

5.1.1 Identity Operator

This operator results in the creation of an output image which is identical to the input image, as shown in Figure 5-3. The value of each pixel in the second image q is identical to the value of the corresponding pixel p in the first image. The function f is a straight line starting at the origin and extending to the maximum pixel value of the image system. The procedure is basic to image processing and other operations are accomplished by modifying the function.

5.1.2 Inverse Operator

This operator results in the creation of an output image which is the inverse of the input image. The process is similar to the identity operator except the function value is different, as shown by Figure 5-4. The function f is a straight line having a value of maximum gray value at the minimum gray input value and equal to zero at the maximum gray input value.

5.1.3 Threshold Operator

This class operator results in a binary output image from a gray scale input image where the level of transition is given by the input parameter p_1 as shown in Figure 5-5 and is known as the threshold. All pixel values below p_1 are converted to a zero and all pixel values equal to or greater than p_1 are converted to a one. This operator can be used to obtain spatial information from the histogram by repeating the procedure using different threshold values.

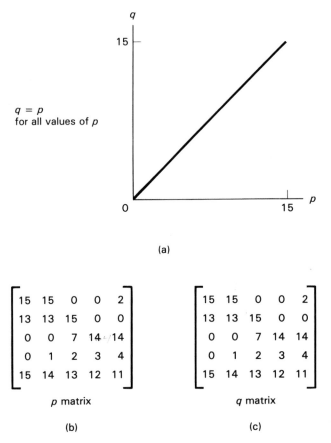

Figure 5-3   Identity operator: function (a), input (b), and output (c).

5.1.4 Other Operations

Different operators can be developed by combining the three basic operators, identity, inverse, and threshold.

Inverted threshold operator. The binary output image can be inverted by applying the threshold function in Figure 5-6 to the image. The threshold value in Figure 5-6 is indicated by p_1, and the example uses a threshold of 5. The point indicates that the value is included. Every pixel in the original image which was light will be dark and pixels which were dark will be light. This operator has extreme value because the threshold value can be varied to obtain spacial information.

Binary threshold interval. This class of operators results in a binary output image where the all input gray values in the interval p_1 to p_2 including p_1 and p_2 are converted to 1 and all input values outside the interval p_1 to p_2

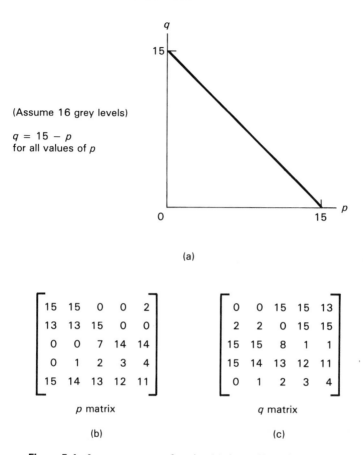

(a)

$$
\begin{bmatrix}
15 & 15 & 0 & 0 & 2 \\
13 & 13 & 15 & 0 & 0 \\
0 & 0 & 7 & 14 & 14 \\
0 & 1 & 2 & 3 & 4 \\
15 & 14 & 13 & 12 & 11
\end{bmatrix}
$$

p matrix

(b)

$$
\begin{bmatrix}
0 & 0 & 15 & 15 & 13 \\
2 & 2 & 0 & 15 & 15 \\
15 & 15 & 8 & 1 & 1 \\
15 & 14 & 13 & 12 & 11 \\
0 & 1 & 2 & 3 & 4
\end{bmatrix}
$$

q matrix

(c)

Figure 5-4   Inverse operator: function (a), input (b), and output (c).

are converted to 0, as illustrated by Figure 5-7. The operator can be used to obtain spacial information from the histogram because it indicates the location of pixels having a given range of values.

Inverted binary threshold operator.   This operator can be used to convert a multilevel gray scale image to a binary image, or it can be applied to a binary image for inversion purposes.

The binary output image, or image produced by the binary threshold interval operator, can be inverted by the inverted binary threshold function as shown by Figure 5-8.

Gray scale threshold operator.   The class of operator shown in Figure 5-9 result in a gray scale output image for gray scale image values between p_1 and p_2 and makes all input values outside the p_1 to p_2 zero. This operator

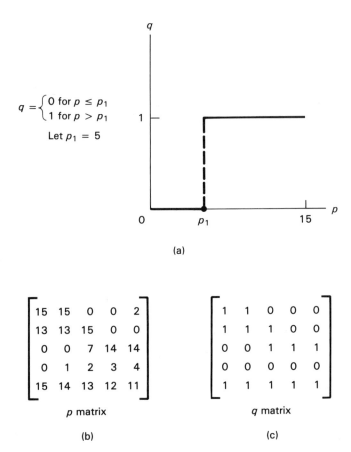

$$q = \begin{cases} 0 \text{ for } p \le p_1 \\ 1 \text{ for } p > p_1 \end{cases}$$

Let $p_1 = 5$

(a)

$$p \text{ matrix} = \begin{bmatrix} 15 & 15 & 0 & 0 & 2 \\ 13 & 13 & 15 & 0 & 0 \\ 0 & 0 & 7 & 14 & 14 \\ 0 & 1 & 2 & 3 & 4 \\ 15 & 14 & 13 & 12 & 11 \end{bmatrix}$$

(b)

$$q \text{ matrix} = \begin{bmatrix} 1 & 1 & 0 & 0 & 0 \\ 1 & 1 & 1 & 0 & 0 \\ 0 & 0 & 1 & 1 & 1 \\ 0 & 0 & 0 & 0 & 0 \\ 1 & 1 & 1 & 1 & 1 \end{bmatrix}$$

(c)

Figure 5-5   Threshold operator: function (a), input (b), and output (c).

can be used to identify image features having a specific value for pseudocolor application processing techniques.

Inverted gray scale threshold operator.   The output image can be inverted using the function illustrated by Figure 5-10. This operator can highlight specific features such as roads or all the similar areas in the image.

Stretch operator.   This class of operators results in a full gray scale output image corresponding to the input interval p_1 and p_2 and suppresses all values outside this range. This is shown in Figure 5-11.

Gray level reduction operator.   This class of operators results in an output image which has a smaller number of gray levels than the number of gray levels of the input image, as shown in Figure 5-12, where an input image with 15 levels is converted to an image having five levels: $q_1, q_2, q_3, q_4, 15$.

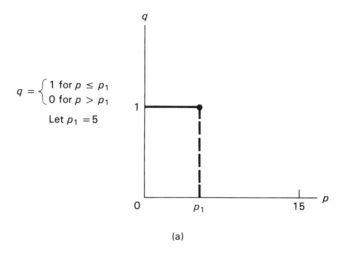

$$q = \begin{cases} 1 \text{ for } p \leq p_1 \\ 0 \text{ for } p > p_1 \end{cases}$$

Let $p_1 = 5$

(a)

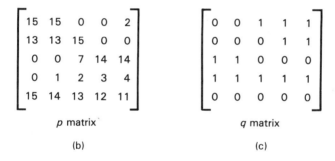

$$\begin{bmatrix} 15 & 15 & 0 & 0 & 2 \\ 13 & 13 & 15 & 0 & 0 \\ 0 & 0 & 7 & 14 & 14 \\ 0 & 1 & 2 & 3 & 4 \\ 15 & 14 & 13 & 12 & 11 \end{bmatrix} \qquad \begin{bmatrix} 0 & 0 & 1 & 1 & 1 \\ 0 & 0 & 0 & 1 & 1 \\ 1 & 1 & 0 & 0 & 0 \\ 1 & 1 & 1 & 1 & 1 \\ 0 & 0 & 0 & 0 & 0 \end{bmatrix}$$

p matrix   q matrix

(b)   (c)

Figure 5-6 Inverted threshold operator: function (a), input (b), and output (c).

5.2 DYADIC TWO POINT TRANSFORMATIONS

The dyadic point-by-point operator uses the information contained at the same location in two images. In the two-input image matrix in Figure 5-13, pixels from A and B are used to create a new image, C. The size of the matrix does not change and the dyadic transformation function f_D can be either linear or nonlinear.

The transformation function is applied to all pixel location pairs in the input images. That is, the information from the pixel location in one image is combined with the information of the corresponding pixel location of a second image to produce the value of a corresponding pixel in the third image. The characteristic function is given by the equation

$$c_{i,j} = f_D(a_{i,j}, b_{i,j})$$

where f_D is a function of two variables and i and j range from 0 to m and n,

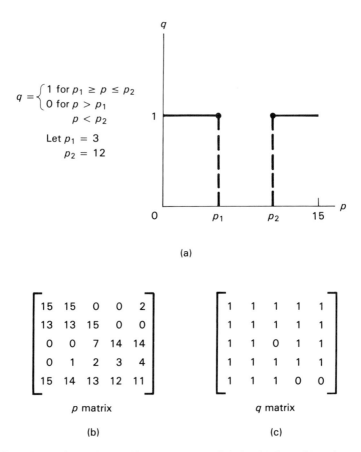

$$q = \begin{cases} 1 \text{ for } p_1 \geq p \leq p_2 \\ 0 \text{ for } p > p_1 \\ \quad\quad p < p_2 \end{cases}$$

Let $p_1 = 3$
$p_2 = 12$

(a)

$$\begin{bmatrix} 15 & 15 & 0 & 0 & 2 \\ 13 & 13 & 15 & 0 & 0 \\ 0 & 0 & 7 & 14 & 14 \\ 0 & 1 & 2 & 3 & 4 \\ 15 & 14 & 13 & 12 & 11 \end{bmatrix}$$

p matrix

(b)

$$\begin{bmatrix} 1 & 1 & 1 & 1 & 1 \\ 1 & 1 & 1 & 1 & 1 \\ 1 & 1 & 0 & 1 & 1 \\ 1 & 1 & 1 & 1 & 1 \\ 1 & 1 & 1 & 0 & 0 \end{bmatrix}$$

q matrix

(c)

Figure 5-7   Binary threshold interval operator: function (a), input (b), and output (c).

respectively. The f_D can be addition, subtraction, multiplication, division, exponentiation, maximum, or any other function that can be devised. Care must be taken to insure that the images are stationary and that there is proper calibration and **registration** of the two input images since the calculations are based on corresponding point by point determination.

The function should contain an appropriate scaling factor k to keep the magnitude of the output values within the scale range to avoid an overflow or a negative condition.

Transformation involves two variables associated with the pairs of corresponding pixels,

$$R(i,j) = f[p(i,j), q(i,j)]$$

where p and q are input matrices, f is the functional operator, and R is the resultant output matrix.

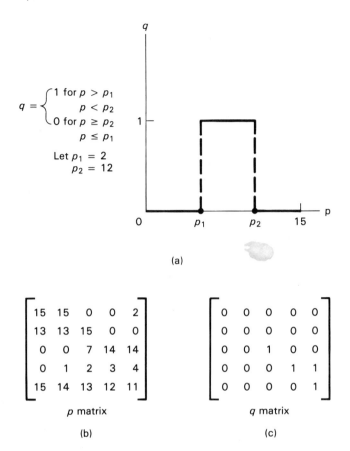

$$q = \begin{cases} 1 \text{ for } p > p_1 \\ \phantom{1 \text{ for }} p < p_2 \\ 0 \text{ for } p \geq p_2 \\ \phantom{0 \text{ for }} p \leq p_1 \end{cases}$$

Let $p_1 = 2$
$p_2 = 12$

(a)

$$\begin{bmatrix} 15 & 15 & 0 & 0 & 2 \\ 13 & 13 & 15 & 0 & 0 \\ 0 & 0 & 7 & 14 & 14 \\ 0 & 1 & 2 & 3 & 4 \\ 15 & 14 & 13 & 12 & 11 \end{bmatrix}$$

p matrix

(b)

$$\begin{bmatrix} 0 & 0 & 0 & 0 & 0 \\ 0 & 0 & 0 & 0 & 0 \\ 0 & 0 & 1 & 0 & 0 \\ 0 & 0 & 0 & 1 & 1 \\ 0 & 0 & 0 & 0 & 1 \end{bmatrix}$$

q matrix

(c)

Figure 5-8   Inverted binary threshold operator: function (a), input (b), and output (c).

5.2.1 Image Addition

Image addition can be used to reduce the effects of noise in the data, as shown in Figure 5-14. The value of the output $C_{i,j}$ is given by

$$c_{i,j} = (a_{i,j} + b_{i,j}) / k$$

over the i, j range of values where k equals the number of samples. The image addition dyadic process averages the data in the two input image matrices. If one of the input images is a constant, the result will be a lighter or a darker overall image which will appear as a shift in the histogram.

In the example, special rules, like rounding up, must be devised and applied to the transformation function to produce meaningful results. There is an accuracy improvement and reduction of noise when the procedure is used with a large number of samples.

$$c_{i,j} = \frac{(a_{i,j} + b_{i,j})}{2}$$

over the i, j range of values. Values are rounded upward.

5.2.2 Image Subtraction

Image subtraction can be used to detect changes that have occurred during the time interval between when two images were taken if the two images are of the same scene. The data may also represent heat losses or cooling if the infrared spectrum is a data source. Since image processing uses positive numbers, it is necessary to define the output in some manner that makes all the values positive. This could involve a rescaling where the largest negative number is set equal to zero and the largest number is set to the maximum gray scale value (255 in part a of Figure 5-15). Another approach is to define the difference as an unsigned difference or the absolute value of the difference as shown in matrix b.

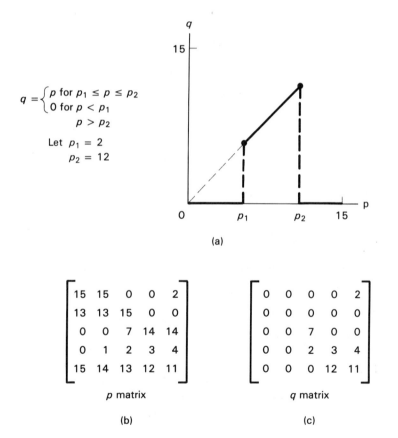

$$q = \begin{cases} p \text{ for } p_1 \leq p \leq p_2 \\ 0 \text{ for } p < p_1 \\ p > p_2 \end{cases}$$

Let $p_1 = 2$
$p_2 = 12$

(a)

$$
\begin{bmatrix}
15 & 15 & 0 & 0 & 2 \\
13 & 13 & 15 & 0 & 0 \\
0 & 0 & 7 & 14 & 14 \\
0 & 1 & 2 & 3 & 4 \\
15 & 14 & 13 & 12 & 11
\end{bmatrix}
$$

p matrix

(b)

$$
\begin{bmatrix}
0 & 0 & 0 & 0 & 2 \\
0 & 0 & 0 & 0 & 0 \\
0 & 0 & 7 & 0 & 0 \\
0 & 0 & 2 & 3 & 4 \\
0 & 0 & 0 & 12 & 11
\end{bmatrix}
$$

q matrix

(c)

Figure 5-9   Gray scale threshold operator: function (a), input (b), and output (c).

In the first case (Example a), the relationship is given by
$$c_{i,j} = k\,(a_{i,j} - b_{i,j})$$
where k is a nonlinear function such that the minimum value of $c_{i,j}$ is 0 and the maximum is 255.

In Example b of Figure 5-15, the relationship is given by
$$R_{i,j} = k\,(a_{i,j} - b_{i,j})$$
where $k = 1$, since the values range from 0 to 254.

The outputs of the two methods are completely different. The system designer or researcher must determine which method provides the most meaningful results for the specific application.

The two examples demonstrate that there are many different ways that the basic processes of addition, subtraction, and multiplication can be used to enhance the data in an image matrix. The designer must determine the type of

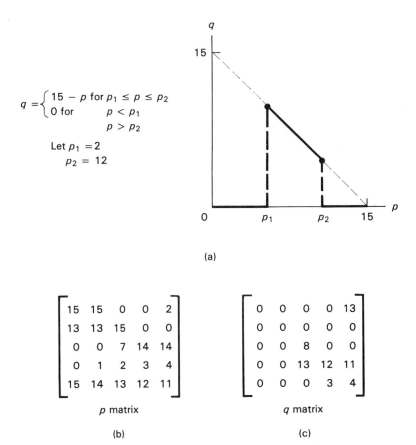

$$q = \begin{cases} 15 - p & \text{for } p_1 \le p \le p_2 \\ 0 & \text{for } \quad p < p_1 \\ & \qquad\; p > p_2 \end{cases}$$

Let $p_1 = 2$
$p_2 = 12$

(a)

$$p \text{ matrix} \quad
\begin{bmatrix}
15 & 15 & 0 & 0 & 2 \\
13 & 13 & 15 & 0 & 0 \\
0 & 0 & 7 & 14 & 14 \\
0 & 1 & 2 & 3 & 4 \\
15 & 14 & 13 & 12 & 11
\end{bmatrix}$$

(b)

$$q \text{ matrix} \quad
\begin{bmatrix}
0 & 0 & 0 & 0 & 13 \\
0 & 0 & 0 & 0 & 0 \\
0 & 0 & 8 & 0 & 0 \\
0 & 0 & 13 & 12 & 11 \\
0 & 0 & 0 & 3 & 4
\end{bmatrix}$$

(c)

Figure 5-10   Inverted gray scale threshold operator: function (a), input (b), and output (c).

information required for the application and the data processing techniques that will enhance the information features in the image data. The monadic operator f will be applied to $c_{i,j}$ matrix in Figure 5-16. $R_{i,j}$ values are rounded up. The equation

$$R_{i,j} = (c_{i,j} + 100) \times \frac{255}{354}$$

is used to convert the values. $R_{i,j}$ will be 0 when $c_{i,j}$ is -100 (minimum value) and 255 when $c_{i,j}$ is 254 (maximum value).

5.2.3 Image Multiplication

The multiplication of two matrixes in image processing is shown in Figure 5-17. This is used to correct for nonlinearity of the sensor where there is a spacial nonuniform sensitivity over the viewing area. This would be corrected by multiplying the image matrix by a correction matrix. The relationship is given by the equation

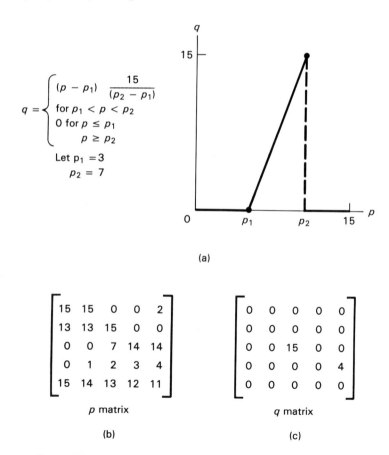

$$q = \begin{cases} (p - p_1) \dfrac{15}{(p_2 - p_1)} \\ \text{for } p_1 < p < p_2 \\ 0 \text{ for } p \le p_1 \\ \phantom{0 \text{ for }} p \ge p_2 \end{cases}$$

Let $p_1 = 3$
$p_2 = 7$

(a)

p matrix

$$\begin{bmatrix} 15 & 15 & 0 & 0 & 2 \\ 13 & 13 & 15 & 0 & 0 \\ 0 & 0 & 7 & 14 & 14 \\ 0 & 1 & 2 & 3 & 4 \\ 15 & 14 & 13 & 12 & 11 \end{bmatrix}$$

(b)

q matrix

$$\begin{bmatrix} 0 & 0 & 0 & 0 & 0 \\ 0 & 0 & 0 & 0 & 0 \\ 0 & 0 & 15 & 0 & 0 \\ 0 & 0 & 0 & 0 & 4 \\ 0 & 0 & 0 & 0 & 0 \end{bmatrix}$$

(c)

Figure 5-11 Stretch operator: function (a), input (b), and output (c).

$$c_{i,j} = k\,[(a_{i,j} \times b_{i,j}) + a_{i,j}]$$

where all values are rounded up to next integer, the maximum value is 255, and $b_{i,j}$ is the correction factor.

Another use for the multiplication operator would be to create a **window** in order to reduce the computation to a specific area of interest. The correction matrix $b_{i,j}$ would be given by the equation

$$c_{i,j} = a_{i,j} \times b_{i,j}$$

where $b_{i,j}$ is 1 for all pairs inside the window area and 0 for all pairs outside the window area.

The size of the output image matrix is identical to the size of the input matrix. No rows or columns are lost in the process as happens with convolutions. The dyadic process for transformation is used in image addition, image subtraction, and image multiplication. A normalizing constant may be applied during the dyadic operation or as a nomadic scaling operator function in a subsequent phase.

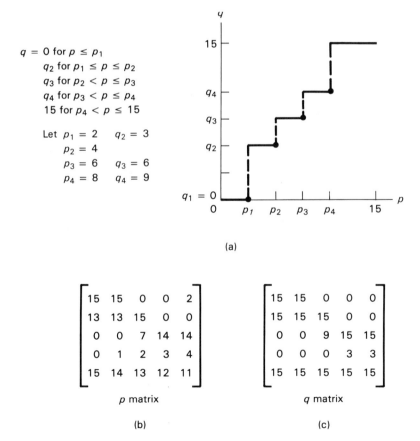

(a)

(b)   (c)

Figure 5-12   Level reduction operator: function (a), input (b), and output (c).

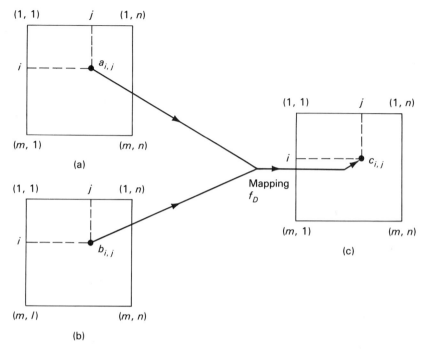

Figure 5-13 Dyadic multipoint-to-point operator: input image A (a) and input image B (b) are mapped into a new output image C (c).

$$
\begin{bmatrix}
0 & 12 & 142 & 255 \\
1 & 6 & 40 & 254 \\
24 & 0 & 20 & 255 \\
30 & 2 & 10 & 240
\end{bmatrix}
\quad
\begin{bmatrix}
14 & 11 & 9 & 253 \\
3 & 5 & 39 & 254 \\
11 & 1 & 19 & 255 \\
18 & 2 & 11 & 256
\end{bmatrix}
\quad
\begin{bmatrix}
7 & 12 & 76 & 254 \\
2 & 6 & 40 & 254 \\
18 & 1 & 20 & 255 \\
23 & 2 & 11 & 248
\end{bmatrix}
$$

$a_{i,j}$	$b_{i,j}$	$c_{i,j}$
input 1	input 2	output

Figure 5-14 Image addition for $i=j=4$.

(a)

$$
\begin{bmatrix}
0 & 12 & 142 & 255 \\
1 & 6 & 40 & 254 \\
24 & 0 & 20 & 255 \\
30 & 2 & 10 & 240
\end{bmatrix}
\quad
\begin{bmatrix}
14 & 11 & 9 & 253 \\
3 & 25 & 100 & 0 \\
11 & 1 & 80 & 1 \\
30 & 2 & 110 & 255
\end{bmatrix}
\quad
\begin{bmatrix}
-14 & 3 & 133 & 2 \\
-2 & -19 & -60 & 254 \\
15 & -1 & -60 & 254 \\
0 & 0 & -100 & -15
\end{bmatrix}
$$

Input #1 = $a_{i,j}$	Input #2 = $b_{i,j}$	Output = $c_{i,j}$ *

(b)

$$
\begin{bmatrix}
14 & 3 & 133 & 2 \\
2 & 19 & 60 & 254 \\
15 & 1 & 60 & 254 \\
0 & 0 & 100 & 15
\end{bmatrix}
$$

Output $c_{i,j}$

Figure 5-15 Image subtraction (absolute value); simple, but useful only under very limited conditions.

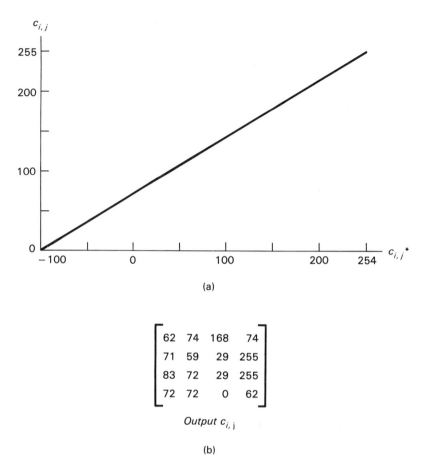

(a)

$$\begin{bmatrix} 62 & 74 & 168 & 74 \\ 71 & 59 & 29 & 255 \\ 83 & 72 & 29 & 255 \\ 72 & 72 & 0 & 62 \end{bmatrix}$$

Output $c_{i,j}$

(b)

Figure 5-16   Image subtraction with function (a) which generates resultant output (b) having all positive values.

5.3 CONVOLUTION: SPATIAL TRANSFORMATION

A new image matrix can be generated where the pixel value assigned to each location is a function of the pixel values of the adjacent locations as is shown by the 3 × 3 convolution in Figure 5-18.

In a 3 × 3 convolution the pixel value of the center location is computed

$$\begin{bmatrix} 0 & 12 & 142 & 255 \\ 1 & 6 & 40 & 254 \\ 24 & 0 & 20 & 255 \\ 30 & 2 & 10 & 240 \end{bmatrix} \quad \begin{bmatrix} .3 & .4 & .1 & .1 \\ .3 & 0 & 0 & .1 \\ .3 & 0 & 0 & 0 \\ .4 & .1 & 0 & .1 \end{bmatrix} \quad \begin{bmatrix} 0 & 17 & 157 & 255 \\ 2 & 6 & 40 & 255 \\ 32 & 0 & 20 & 255 \\ 42 & 3 & 10 & 255 \end{bmatrix}$$

Image matrix $a_{i,j}$ Correction matrix $b_{i,j}$ Output matrix $c_{i,j}$

Figure 5-17   Image multiplication.

$$\begin{bmatrix} a & b & c \\ d & e & f \\ g & h & i \end{bmatrix} \times \begin{bmatrix} + & + & - \\ - & + & + \\ + & + & - \end{bmatrix} = \begin{bmatrix} \boxed{e'} \end{bmatrix}$$

Where:

$$e' = + a + b - c - d + e + f + g + h - i$$

Figure 5-18 Convolution (neighborhood) transformation.

based on the values of the nine locations. The location is then shifted by 1 and the process repeated until the entire image matrix is generated. It should be noted that the size of the resultant matrix is reduced by 2 in each direction due to the edge effect. If it is undesirable to reduce the matrix size, this can be modified using a rule which assumes the data is constant in the row and column adjacent to the outside edge of the image matrix.

REFERENCES

1. Diana Ngyssonen, "Practical Method for Edge Detection and Focusing for Linewidth Measurements on Wafers, *Optical Engineering,* January, 1987.
2. Andrew C. Wilson, Sr., "Imaging Joins Forces with AI," *Electronic Systems Design Magazine,* January, 1987.
3. Mike Kayat, "Array Processor Spurs Faster Imaging," *Electronic Systems Design Magazine,* February, 1987.
4. R. C. Gonzalez, R. E. Woods, & W. T. Swain, "Digital Image Processing: An Introduction," *Digital Design,* March, 1986.

EXERCISES

1. Given 6 × 6 image matrix p
   (a) Draw a monadic function which will result in an output matrix q which stretches p values between 3 to 11 into q values ranging between 0 and 255. All other p values outside the 3 to 11 interval map into zero gray level.
   (b) Determine the output matrix when the transfer matrix in Figure 5-20 is applied to the image data in Figure 5-19.

$$\begin{bmatrix} 0 & 4 & 8 & 12 & 12 & 15 \\ 2 & 2 & 4 & 5 & 6 & 8 \\ 4 & 1 & 5 & 10 & 12 & 14 \\ 5 & 3 & 5 & 6 & 8 & 8 \\ 8 & 4 & 6 & 8 & 10 & 4 \\ 12 & 13 & 14 & 15 & 0 & 2 \end{bmatrix}$$

Image matrix p

Figure 5-19 Image data.

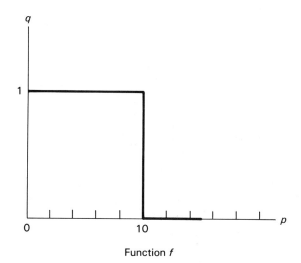

Function *f*

Figure 5-20 Monadic operator function.

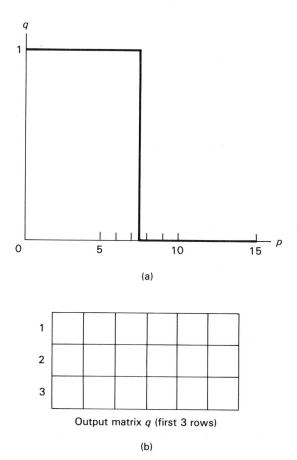

(a)

Output matrix *q* (first 3 rows)

(b)

Figure 5-21 Monadic operator function (a) and output (b).

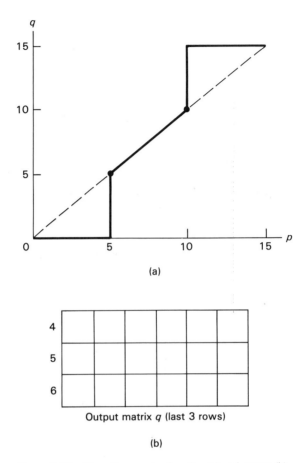

(a)

Output matrix q (last 3 rows)

(b)

Figure 5-22   Monadic operator function (a) and output (b).

2. Give the first three rows of the output matrix q when the function in Figure 5-21 has been applied to input matrix p in Figure 5-19.

3. Give the last three rows of the output matrix q when the function in Figure 5-22 has been applied to the input matrix p in Figure 5-19.

4. Determine the output 6×6 matrix if the function in Figure 5-23 operates on input matrix p in Figure 5-19.

5. Write a computer program which utilizes the input matrix p and each of the functions to produce the output matrix q.

6. Write a computer program which produces a histogram of the input matrix p, of the output matrix q and a histogram of the difference between p and q.

7. Determine the calibration matrix and determine the output matrix $q(i,j)$ which is the result of applying a calibration matrix to the input matrix in Figure 5-19 of a 16 gray level system. The calibration or correction required is based on the center of

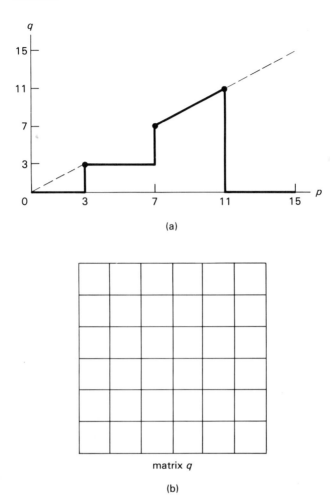

Figure 5-23 Monadic operator function (a) and output (b).

the 6 × 6 array and varies from the center to the outer edge in the following manner:

   −10% for pixels one pixel from the center of the image; that is, the distortion of the lens is such that the values in the matrix are too high,

   0 % for pixels two pixels distant from the center of the image,

   +10% for pixels at a distance greater than 2 pixels from the center of the image.

Use rounding rule that rounds down from 0.5 and up for values ⩽ 0.5.

8. Determine $c(i,j)$ where $c(i,j) = a(i,j) - b(i,j)$ and the minimum value is 0 and the maximum value is 255.
   $a(i,j)$ and $b(i,j)$ are

$$\begin{bmatrix} 0 & 12 & 142 & 255 \\ 1 & 6 & 40 & 254 \\ 24 & 0 & 20 & 255 \\ 30 & 2 & 10 & 240 \end{bmatrix} \quad \begin{bmatrix} 14 & 11 & 9 & 253 \\ 3 & 5 & 39 & 254 \\ 11 & 1 & 19 & 255 \\ 18 & 2 & 11 & 256 \end{bmatrix}$$

$$a_{i,j}$$
input #1

$$b_{i,j}$$
input #2

9. Match the columns.

1. Monadic   _____   a. Function used in process of smoothing image data.

2. Dyadic   _____   b. Value at which some function or process starts to take effect.

3. Convolution   _____   c. New pixel value is generated based on the pixel values of the adjacent locations.

4. Inverse operator   _____   d. New pixel value is generated based on the pixel values of corresponding locations in two or more input images.

5. Threshold   _____   e. Process used to detect changes that have occurred during time interval when two images are acquired.

6. Stretch operator   _____   f. Monadic function which can produce an image which is the inverse of the input image.

7. Image subtraction   _____   g. New pixel value is generated based on the pixel value of the specific location.

8. Image addition   _____   h. Process used to correct systematic error which is a function of the location in the matrix or to produce a window in order to reduce computation to a specific area of interest.

9. Image multiplication   _____   i. Increases or decreases the range of gray scale values.

6

EDGE ENHANCEMENT

6.0 INTRODUCTION

Each system will have a very specific operational purpose and is called upon to make "go" or "no-go" decisions. The purpose varies considerably depending on the function and type of organization. An equipment manufacturer's needs are different from those of a pharmaceutical concern. The purpose will include tasks such as insuring that the product is within dimensional specifications as part of an inspection function, reading bar code information for manufacturing control, identifying different products on a conveyor belt in a sorting operation, or verifying that the correct components were used in an assembly operation.

The system first acquires image data on the specific object or process, which contains information relevant to the parameters involved in making the decision required for accomplishment of the task. Once the data is in the system, monadic and dyadic image processing techniques will be used to enhance the image data. Each pixel can undergo a different transformation because each pixel has a unique local neighborhood. The application of the global operator should work toward reducing the information about the image rather than increasing it. The operator should enhance meaningful relationships and minimize noise effects. For example, if the contrast between light and dark areas is very low, the monadic stretch operator covered in Section 5.1.4 could be applied on a global scale to expand the difference in contrast over a wider scale prior to subsequent processing. The key elements in processing are the ability to systematically manipulate the basic data in an acceptable period of time to make the appropriate decision, and to be certain that image data is always processed in the same manner for reliability of results. The processing operator that can be designed to do different transformations will calculate a new value for each pixel based on the value of the pixels around it. The operational process is referred to as a digital filter.

6.1 DIGITAL FILTER

The digital filter can be categorized as a low pass or a high pass filter depending on which part of the frequency spectrum it affects. The effect on a composite electrical analog signal containing all frequency components is shown in Figure 6-1.

The output signal of the high pass filter will contain the high frequency components. The output signal of the low pass filter will contain the DC bias and the low frequency components. The specific cutoff frequency can be varied by the selection of the values used in the filter, but that is beyond the scope of this course.

The low frequency components in an image are characterized by a slow change in the contrast or values of adjacent pixels in relation to high frequency components, due to the rapid change in contrast encountered at the edge of an object.

6.2 LOW PASS FILTER

The low pass filter will not affect low frequency components in the image data and will attenuate the high frequency components as illustrated in Figure 6-2. Random speckles in an image can be considered as noise and are high

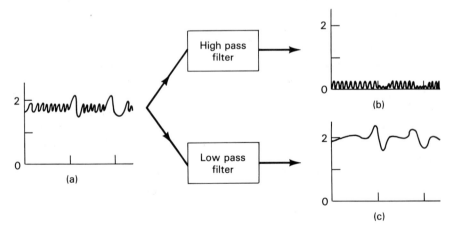

Figure 6-1   Results of low pass (b) and high pass (c) digital filters on a composite signal (a) containing both low and high frequencies as well as a DC bias.

$$\begin{bmatrix} f_{1,1} & f_{2,1} & f_{3,1} \\ f_{1,2} & f_{2,2} & f_{3,2} \\ f_{1,3} & f_{2,3} & f_{3,3} \end{bmatrix} = \begin{bmatrix} 1/9 & 1/9 & 1/9 \\ 1/9 & 1/9 & 1/9 \\ 1/9 & 1/9 & 1/9 \end{bmatrix}$$

Figure 6-2   Coefficient values of a low pass filter matrix.

frequency components since the pixel values of adjacent pixels change very rapidly. The effect of the speckles can be reduced by using a simple averaging filter. The coefficient matrix for an averaging filter for a 3 × 3 local area **convolution** would contain nine elements as shown below.

The value for the pixel in the output matrix is calculated by convoluting the image data, $p_{i,j}$ with the coefficient matrix $f_{i,j}$ and placing the resultant value at the center pixel of the area in the new image matrix. The new value of $q_{2,2}$ is given by

$$q_{2,2} = f_{1,1} \times p_{1,1} + f_{2,1} \times p_{2,1} + \ldots + f_{3,3} \times p_{3,3}$$

The rules for a low pass filter are

1. All the coefficients must be positive, and

2. The sum of all the coefficients must equal 1.

If the sum of the coefficients is greater than 1, amplification will result. If the sum is less than 1, attenuation will result, and the magnitude of the pixel values over the entire image will be reduced. Hence, the image would appear darker.

The application of the 3 × 3 coefficient matrix to a data field where all the pixels have the same value will have no effect except where there is a gradient between adjacent pixels. The effect of the low pass digital filter in reducing noise can be illustrated by considering the case where there is one pixel in the matrix which is zero due to a bad element in the sensor array used to acquire the data; all the adjacent pixels have a high value.

$$\begin{bmatrix} 7 & 7 & 7 & 7 \\ 7 & ⓪ & 7 & 7 \\ 7 & 7 & 7 & 7 \end{bmatrix} \qquad \begin{bmatrix} 7 & 7 & 7 & 7 \\ 7 & ⑦ & 7 & 7 \\ 7 & 7 & 7 & 7 \end{bmatrix}$$

matrix with   matrix after
bad element   low pass filter

If the bad pixel value had been high instead of low as the result of a white speck, the application of the low pass filter would have had the effect of reducing the magnitude of the value of the bad pixel.

The effect of increasing the size of the neighborhood region 5 × 5 or 9 × 9 can be illustrated by comparing the results of applying a low pass filter to the case of one bad pixel value, (0 in a region contains all ones) as is done in Table 6-1.

It can be seen that the larger size matrix causes a new calculated value which approaches the value of the adjacent pixels but at a greatly increased nine-fold cost of computational time. The 3 × 3 is commonly used for most

TABLE 6-1   Effect of Filter Dimensions

Size of Matrix	Convoluted Value	Raw Value of Bad Element	Relative Computation Time
3×3	1/9	8/9	1
5×5	1/25	24/25	$25/9 = 2\text{-}7/9$
9×9	1/81	80/81	$81/9 = 9$

industrial applications, as it greatly reduces the computational processing requirements, the processing time, and equipment cost; the results should be checked to insure they are within the acceptable error tolerance requirements.

6.3 HIGH PASS FILTER

The high pass digital filter has the inverse characteristic of the low pass filter. The filter will not change the high frequency components of a signal and will attenuate the low frequency components as well as eliminate any bias in the signal.

The effect of background light would result in a DC bias in the vision system data since the minimum value of all the pixels would be above some given value. The effect of applying a high pass filter to the data for a 6×6 image is shown in the histograms of Figure 6-3. The high pixel values may or may not change depending on the degree of saturation of the image.

A high-pass filter coefficient set for a 3×3 matrix convolution mask is

$$\begin{bmatrix} -1 & -1 & -1 \\ -1 & 8 & -1 \\ -1 & -1 & -1 \end{bmatrix}$$

1. The coefficients can be positive or negative.
2. The sum of the coefficients is 0.

Since the sum of all the coefficients is equal to zero, the DC component will be completely suppressed. If the sum were equal to $+1$, the DC value would have been retained at the same value as the original signal. The filter is applied by overlaying the 3×3 coefficient matrix filter mask on the image data, multiplying the coefficient and data in each of the nine elements, and adding the nine resultant products to obtain the value of the new element located at the center of the 3×3 matrix. The mask is then shifted by one unit and the process repeated to obtain the value of the next new element. The following example demonstrates the process to obtain the value in Figure 6-4. Convolution of mask and data gives value of central element at (2,2).

$$\begin{bmatrix} +1 & +1 & +1 \\ +1 & -8 & +1 \\ +1 & +1 & +1 \end{bmatrix} \quad \begin{bmatrix} 4 & 8 & 8 \\ 4 & 8 & 8 \\ 4 & 8 & 8 \end{bmatrix} \quad \begin{bmatrix} - & - & - \\ - & [\ ] & - \\ - & - & - \end{bmatrix}$$

mask   data   element

$$(2,2) = +(1\times4) +(1\times8) +(1\times8) +(1\times4) -(8\times8)$$
$$+(1\times8) +(1\times4) +(1\times8) +(1\times8)$$

$$= +4 +8 +8 +4 -64 +8 +4 +8 +8$$

$$= +52 -64$$

$$= -12$$

An edge in an image matrix will appear as a sharp change in the values of adjacent pixels as shown in Figure 6-4, Example B. The information relative to the location and magnitude of the edge is contained or represented by the high spatial frequency components of the pixel values in the image.

It can be seen that the high pass filter eliminated the constant value bias in Example A and enhanced the edge pixels in Example B where the resultant matrix has a line of low and high pixel values.

The coefficient matrix can be modified slightly where the sum is equal to one, and a different result is achieved when the filter is applied to the data in Examples A and B.

$$\begin{bmatrix} -1 & -1 & -1 \\ -1 & +9 & -1 \\ -1 & -1 & -1 \end{bmatrix}$$

modified
coefficient matrix

$$\begin{bmatrix} x & x & x & x & x & x & x \\ x & 4 & 4 & 4 & 4 & 4 & x \\ x & 4 & 4 & 4 & 4 & 4 & x \\ x & 4 & 4 & 4 & 4 & 4 & x \\ x & x & x & x & x & x & x \end{bmatrix} \quad \begin{bmatrix} x & x & x & x & x & x & x \\ x & 4 & -8 & +20 & 8 & 8 & x \\ x & 4 & -8 & +20 & 8 & 8 & x \\ x & 4 & -8 & +20 & 8 & 8 & x \\ x & x & x & x & x & x & x \end{bmatrix}$$

effect of modified high pass filter on examples a and b.

The result of the convolution of the modified high pass filter with the data in images A and B provides images where the bias is retained and where the high frequency spatial component is amplified.

The 3 × 3 matrix is the smallest symmetrical unit and minimizes the calculation necessary to perform the convolution. The method is easily extended to 5 × 5 and 9 × 9 matrices, but results may not be markedly different while the calculations have been expanded by factors of 2.7 and 9, respectively. The 3 × 3 convolution capability is sufficient for most industrial

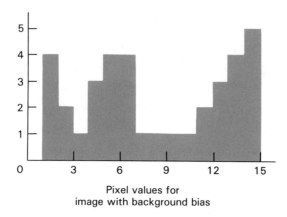

Pixel values for
image with background bias

(a)

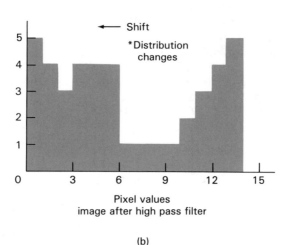

Pixel values
image after high pass filter

(b)

Figure 6-3   Effect of a high pass filter on a histogram. When values of pixels are
shifted, the shape of the histogram changes.

applications. The 9 × 9 result can be achieved on a 3 × 3 system by the
repeated application of the 3 × 3 convolution. The outside values are lost
since there is insufficient data. The high pass filter has the property that the
data in the edge region is modified and the edge effect amplified. **Edge
detection** is basic to obtaining knowledge from an image because the informational content is related to the concepts of contrast, shape, location, and
dimensions.

Edges are the critical parameter to all the concepts and the location of
edges is of paramount importance. The machine vision system uses different
forms of high pass filters to enhance the features of interest.

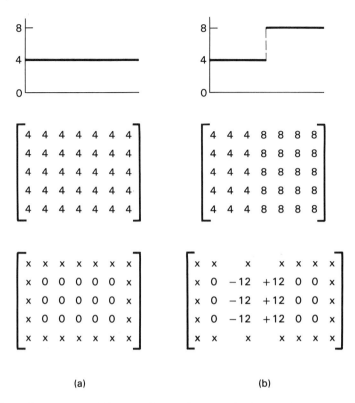

(a)   (b)

Figure 6-4   Results of high pass filter on image not having an edge (a) and on an image containing an edge (b). *x* equals last value because of incomplete data.

6.4 EDGE ENHANCEMENT

An edge is a boundary between two regions of different gray levels. The ideal edge is a step function and can be sharp or diffuse. The diffuse or wide edge can be considered to consist of a number of closely located small edges, as is shown in Figure 6-5. In the real world of machine vision applications, the edge will usually have noise associated with the data. High spatial frequencies are associated with sharp changes in the intensity values of the pixels. The enhancement of edges can be accomplished by applying some form of high pass filter to the image data. A large number of edge enhancement operators exist in the literature, but the most popular in the industrial environment are

Laplacian edge enhancement
Robert's gradient
Sobel edge detector

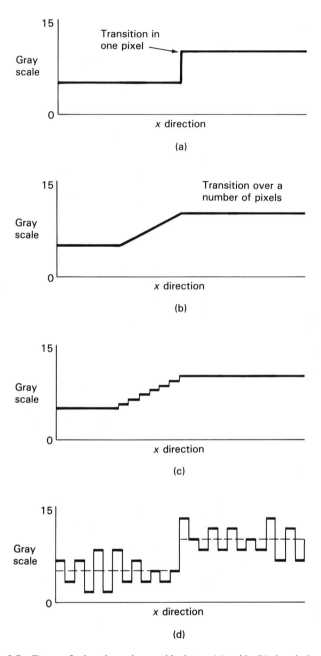

Figure 6-5 Types of edges in an image: ideal step (a), wide (b), break down of wide edge (c), and ideal step with noise (d).

A step-type edge separates two regions with relatively uniform gray levels on each side of the edge. A line should be considered to have two step edges, one positive and one negative. The width of the line is dependent on the separation distance between the two steps.

6.4.1 Laplacian Edge Enhancement Operator

The **Laplacian operator** used in image processing is based on the mathematical second partial derivative Laplacian expression in continuous functions.

$$\nabla^2 f = \frac{\partial^2 f}{\partial x^2} + \frac{\partial^2 f}{\partial y^2}$$

This continuous mathematical expression can be approximated by the difference operators in the discrete matching vision digital processing application. The **rotationally insensitive** Laplacian operator process is essentially determining the *change in slope* of the intensity at the pixel in the x and y directions.

$$L(i,j) = \nabla^2 p(i,j) = \Delta x^2 p(i,j) + \Delta y^2 p(i,j)$$

where

$$\Delta x^2 = [p(i-1, j) - p(i,j)] - [p(i,j) - p(i+1, j)]$$
$$\Delta y^2 = [p(i,j+1) - p(i,j)] - [p(i,j) - p(i,j-1)]$$

$$\begin{bmatrix} a & b & c \\ d & e & f \\ g & h & i \end{bmatrix} \qquad \Delta x^2 = [d - e] - [e - f]$$

$$\begin{bmatrix} a & b & c \\ d & e & f \\ g & h & i \end{bmatrix} \qquad \Delta y^2 = [h - e] - [e - b]$$

The collection of terms will result in the following for the Laplacian operator.

$$L(i,j) = b + d + f + h - 4e$$

The $L(i,j)$ can be reduced to the mask

$$\begin{bmatrix} 0 & +1 & 0 \\ +1 & -4 & +1 \\ 0 & +1 & 0 \end{bmatrix}$$

In summary, the Laplacian operator computes the difference between the gray level of the center pixel and the average of the gray levels of the four adjacent pixels in the horizontal and vertical directions. It is a high pass filter because the sum of the coefficients is zero, and it contains both positive and negative coefficients. The application of the Laplacian to the input matrix of Example B containing the vertical edge is shown in Figure 6-6.

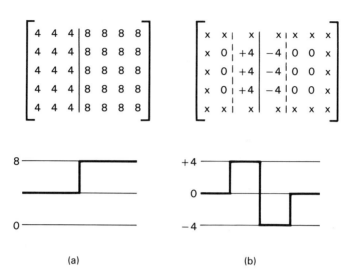

(a)   (b)

Figure 6-6 Laplacian operator applied to image containing a verticle edge (a) results in zero crossing identification of edge in output (b).

The application of the Laplacian to an image with an edge at an angle is shown in Figure 6-7.

The presence of a positive edge as shown in Figure 6-8 results in a positive impulse followed by a negative impulse since the Laplacian operator is differentiating the surface. The edge is located between the two impulse functions at the point where the impulse functions cross zero. Hence, the Laplacian operator can be used to enhance the edge feature, and a zero crossing detector can be used to locate the edge feature. The impulse functions from the four types of edges is shown in Figure 6-8.

The edges and lines in an image can run in any direction. It is, therefore, highly desirable that an operator be **isotropic** if it is to be used to enhance images in blurred features in an image. In order to be isotropic, the operator must involve only derivatives of an even order. The results obtained from

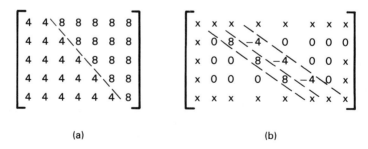

(a)   (b)

Figure 6-7 Laplacian operator applied to image containing an edge at an angle (a) results in output (b).

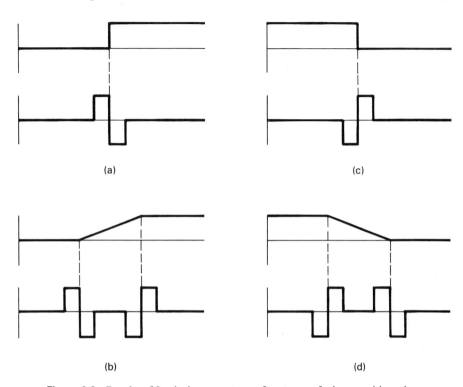

(a)   (c)

(b)   (d)

Figure 6-8 Results of Laplacian operator on four types of edges: positive edge
(a), negative edge (b), positive ramp (c), and negative ramp (d).

applying the isotropic operator to an image and rotating the output are
identical to those obtained when the input is rotated and then the operator is
applied.

The edge enhancement operators usually involve differiention to identify the change in slope of the values associated with the pixels over the
surface. The second derivative may be estimated by the difference of the first
differences.

$$\frac{\partial^2 f}{x^2} = (f_{m,n} - f_{m+1,n}) - (f_{m+1,n} - f_{m+2,n})$$

$$\frac{\partial^2 f}{\partial x^2} = f_{m,n} - 2f_{m+1,n} + f_{m+2,n}$$

Input a, b, c. Approximation of the second derivative in the x direction

$$\frac{\partial^2 f}{\partial x^2} = a - 2b + c$$

$$\begin{vmatrix} \text{Mask} \\ 1 \;\; -2 \;\; 1 \end{vmatrix}$$

The matrix notation which will be used for $f_{i,j}$ is

$$\text{row: } i = 1 \to M \quad (y \text{ direction})$$
$$\text{column: } j = j \to N \quad (x \text{ direction})$$

MN matrix has M rows and N columns

$$\begin{bmatrix} f_{1,1} & \cdots & f_{1,N} \\ \vdots & & \vdots \\ f_{M,1} & \cdots & f_{M,N} \end{bmatrix}$$

6.4.2 Robert's Gradient Operator

The Robert's gradient operator (Figure 6-9) is simpler than the Laplacian operator because it operates on a 2×2 region of pixels at each point. The use of data from this size region rather than the larger 3×3 region of the Laplacian operator reduces the computational time and capability of the system. This operator uses the diagonal derivatives to estimate the gradient at the point. The magnitude operator is equal to the square root of the sum of the squares of the two diagonal differences. It can be approximated using the computationally simpler expression of the sum of the absolute value of each of the diagonal differences.

$$\text{Magnitude operator} = (\Delta1^2 + \Delta2^2)^{1/2}$$

$$\text{Absolute value estimate} = |\Delta1| + |\Delta2|$$

where $\Delta1 = p(i,j) - p(i+1, j+1)$ and $\Delta2 = p(i+1, j) - p(i, j+1)$

$$\begin{bmatrix} a & b \\ c & d \end{bmatrix} \qquad \text{New } q(1,1) = |\Delta1| + |\Delta2|$$

where $\Delta1 = (a-d)$ and $\Delta2 = (b-c)$

$$\begin{bmatrix} 2 & 5 & 6 & 2 \\ 3 & 1 & 3 & 4 \\ 2 & 3 & 5 & 1 \end{bmatrix} \qquad \text{New } q(1,1) = (2-1) + (5-3) = |1| + |2| = 3$$

input

$$\begin{bmatrix} 3 & 7 & 3 & x \\ 1 & 4 & 3 & x \\ x & x & x & x \end{bmatrix}$$

outputs after
Robert's
Operator

Figure 6-9 Robert's gradient operator.

6.4.3 Sobel Edge Detector Operator

The Sobel operator (Figure 6.10) is a nonlinear computation of the edge magnitude at a point, but it does not use the value at the point itself in the calculation. The value at the pixel is given by

$$S = (\Delta x^2 + \Delta y^2)^{1/2}$$

where

$$\begin{bmatrix} a & b & c \\ d & e & f \\ g & h & i \end{bmatrix}$$
input

$$\Delta x = (a + 2d + g) - (c + 2f + e)$$

$$\Delta y = (g + 2h + i) - (a + 2b + c)$$

$$\Delta x = \begin{bmatrix} 1 & 0 & -1 \\ 2 & 0 & -2 \\ 1 & 0 & -1 \end{bmatrix} \qquad \Delta y = \begin{bmatrix} -1 & -2 & -1 \\ 0 & 0 & 0 \\ 1 & 2 & 1 \end{bmatrix}$$

convolution masks for sobel operator

$$\begin{bmatrix} 3 & 4 & 2 & 5 & 1 \\ 2 & ① & 6 & 4 & 2 \\ 3 & 5 & 7 & 1 & 3 \\ 4 & 2 & 5 & 7 & 1 \\ 2 & 5 & 1 & 3 & 2 \end{bmatrix} \qquad \begin{matrix} x & x & x & x & x \\ x & q_{2,2} \\ \\ \\ \end{matrix}$$

input

$q_{2,2} = (\Delta x^2 + \Delta y^2)^{1/2}$ around pixel 2,2
$\Delta x = 3 + 4 + 3 - 2 - 12 - 7 = +11$
$\Delta y = -3 - 8 - 2 + 3 + 10 + 7 = +17$
$q^{2,2} = [(-11)^2 + (+7)^2]^{1/2}$
$= [121 + 49]^{1/2}$
$= (170)^{1/2} \cong 13$

Figure 6-10   Sobel edge detector operator.

6.4.4 Other Local Operators

The enhancement methods presented in the previous sections are very basic and are found in the standard processing package of software routines supplied with most commercial vision systems. If the exact nature of how the image degrading function is known, it is possible to predict in advance how different operators will affect the image. In practice, it may be necessary to experiment with various filters to determine which produce the best results. The convolution masks for a number of different operators provide the reader with insight on the method of devising special filters.

CONVOLUTION MASKS

1. Low pass filters (coefficients positive and added to one)

$$\begin{bmatrix} 0.1 & 0.1 & 0.1 \\ 0.1 & 0.2 & 0.1 \\ 0.1 & 0.1 & 0.1 \end{bmatrix} \qquad \begin{bmatrix} 1/16 & 1/8 & 1/16 \\ 1/8 & 1/4 & 1/8 \\ 1/16 & 1/8 & 1/16 \end{bmatrix}$$

2. High pass filters (coefficients added to zero)

$$\begin{bmatrix} 0 & -1 & 0 \\ -1 & 4 & -1 \\ 0 & -1 & 0 \end{bmatrix} \qquad \begin{bmatrix} 1 & -2 & 1 \\ -2 & 4 & -2 \\ 1 & -2 & 1 \end{bmatrix}$$

3. High pass filter with DC Bias (coefficients added to one)

$$\begin{bmatrix} -1 & -1 & -1 \\ -1 & 9 & -1 \\ -1 & -1 & -1 \end{bmatrix} \qquad \begin{bmatrix} 1 & -2 & 1 \\ -2 & 5 & -2 \\ 1 & -2 & 1 \end{bmatrix}$$

4. Gradient-directional (coefficients added to zero)

$$\begin{bmatrix} 1 & 1 & 1 \\ 1 & -2 & 1 \\ -1 & -1 & -1 \end{bmatrix} \qquad \begin{bmatrix} 1 & 1 & 1 \\ -1 & -2 & 1 \\ -1 & -1 & 1 \end{bmatrix} \qquad \begin{bmatrix} -1 & 1 & 1 \\ -1 & -2 & 1 \\ -1 & 1 & 1 \end{bmatrix}$$
$$\quad\text{N} \qquad\qquad\qquad \text{NE} \qquad\qquad\qquad \text{E}$$

5. Shift and difference filters (coefficients added to zero)

$$\begin{bmatrix} 0 & 0 & 0 \\ -1 & 1 & 0 \\ 0 & 0 & 0 \end{bmatrix} \qquad \begin{bmatrix} 0 & -1 & 0 \\ 0 & 1 & 0 \\ 0 & 0 & 0 \end{bmatrix} \qquad \begin{bmatrix} 0 & 0 & -1 \\ 0 & 1 & 0 \\ 0 & 0 & 0 \end{bmatrix}$$
$$\text{Vertical edges} \qquad \text{Horizontal edges} \qquad \begin{array}{c}\text{Horizontal and}\\ \text{vertical edges}\end{array}$$

6. Blurs

$$\begin{bmatrix} 0 & 0 & 0 \\ 1 & 1 & 1 \\ 0 & 0 & 0 \end{bmatrix} \qquad \begin{bmatrix} 0 & 1 & 0 \\ 0 & 1 & 0 \\ 0 & 1 & 0 \end{bmatrix} \qquad \begin{bmatrix} 1 & 0 & 0 \\ 0 & 1 & 0 \\ 0 & 0 & 1 \end{bmatrix}$$
$$\text{Horizontal} \qquad\qquad \text{Vertical} \qquad\qquad \text{Diagonal}$$

7. Difference filters

$$\begin{bmatrix} 0 & 1 & 0 \\ 0 & 0 & 0 \\ 0 & -1 & 0 \end{bmatrix} \qquad \begin{bmatrix} 0 & 0 & 0 \\ 1 & 0 & -1 \\ 0 & 0 & 0 \end{bmatrix}$$
$$\text{Vertical} \qquad\qquad \text{Horizontal}$$

8. Vertical differentiation filters

Take absolute value difference (7).

9. Horizontal differencing and vertical smoothing filters

$$\begin{bmatrix} 1 & 0 & -1 \\ 1 & 0 & -1 \\ 1 & 0 & -1 \end{bmatrix}$$

10. Laplacian filters (coefficients added to zero)

$$\begin{bmatrix} 0 & -1 & 0 \\ -1 & 4 & -1 \\ 0 & -1 & 0 \end{bmatrix} \qquad \begin{bmatrix} 1 & -2 & 1 \\ -2 & 4 & -2 \\ 1 & -2 & 1 \end{bmatrix}$$

11. Bright region expansion

Maximum of

$$\begin{bmatrix} 1 & 1 & 1 \\ 1 & 1 & 1 \\ 1 & 1 & 1 \end{bmatrix}$$

12. Medium filter (reduce camera noise)	Fifth largest of	$\begin{bmatrix} 1 & 1 & 1 \\ 1 & 1 & 1 \\ 1 & 1 & 1 \end{bmatrix}$		

13. Enhance line segment (coefficients added to zero)

$$\begin{bmatrix} -1 & 2 & -1 \\ -1 & 2 & -1 \\ -1 & 2 & -1 \end{bmatrix} \qquad \begin{bmatrix} -1 & -1 & -1 \\ 2 & 2 & 2 \\ -1 & -1 & -1 \end{bmatrix} \qquad \begin{bmatrix} -1 & -1 & 2 \\ -1 & 2 & -1 \\ 2 & -1 & -1 \end{bmatrix}$$

Vertical   Horizontal   L-R Diagonal

14. Edge detector   Max $(A) - (B)$ where:

$$A = \begin{bmatrix} 1 & 1 & 1 \\ 1 & 1 & 1 \\ 1 & 1 & 1 \end{bmatrix} \qquad B = \begin{bmatrix} 0 & 0 & 0 \\ 0 & 1 & 0 \\ 0 & 0 & 0 \end{bmatrix}$$

15. Line detector

$$q_{2,2} = A - \text{Min}(\text{Min}(\text{Max}(\text{Max}(B))))$$

where:

$$A = \begin{bmatrix} 0 & 0 & 0 \\ 0 & 1 & 0 \\ 0 & 0 & 0 \end{bmatrix} \qquad B = \begin{bmatrix} 1 & 1 & 1 \\ 1 & 1 & 1 \\ 1 & 1 & 1 \end{bmatrix}$$

16. Template matching (Cross correlation)

$$\begin{bmatrix} 0 & 0 & 0 \\ 1 & 1 & 1 \\ 0 & 0 & 0 \end{bmatrix}$$

6.5 IMAGE ANALYSIS

The objective in using image enhancing techniques is to produce an improved version of the input image. The next phase in image processing is that of image analysis, to identify parameters associated with a given feature of an object in the picture. A key element in the task of locating features is that of identifying the location of edges. The process of reducing the image to objects or regions is referred to as **segmentation.** The segmentation operation depends on the ability of the machine vision system to detect the edge or abrupt change in the gray level value associated with the pixels where the edge is located.

There is no single approach to segmentation and there are many different ways of identifying the objects in the image. Thresholding is a very common way to extract the objects.

6.5.1 Thresholding

Thresholding is the process of segregating pixels of different gray level values. In the process, the pixels having a gray scale value at or below a given threshold value are given a zero value and all above the threshold are set at one. This function was covered in Section 5.1.4 and is expressed by

$$q_{i,j} = \begin{cases} 1 \text{ if } p_{i,j} \geq \text{threshold value} \\ 0 \text{ if } p_{i,j} < \text{threshold value} \end{cases}$$

The threshold function operator may be applied on a global scale to the entire image, or different threshold values may be used for different objects and regions of the image. The process is ideally suited for objects which have a characteristic set of gray level values, common to industrial applications.

The difference operation (7 in the list), as well as other operations, could be used for the edge location function. The essential task is that of devising the simplest detector that will function for edges in all the desired directions in order to minimize computations. The difference operator could be modified by incorporating a weighted average feature to detect ramplike edges where the weight is highest (at the centers of two touching regions) and taper off as the distance from the center increases. The weights could be assigned in an inverse order to detect abrupt edges.

The Laplacian operator produces positive and negative pixel values at the edge (Figure 6-8). A zero-crossing detector can be used as an edge location indicator. Each threshold value produces a unique pattern and will result in different output images. The output of the threshold operator is an image, containing patterns associated with the informational content of the image. Once the patterns have been obtained, the next step in image processing is to use a pattern identifier to determine the relevance of the outline. This is accomplished with a pattern matching process.

6.5.2 Pattern Matching

The edge location operation identified points in the image where there was a step in the local pattern of gray level values. The matching of patterns would be relatively easy if the two patterns being matched were identical. In practice, there are problems of noise, distortion, and orientation.

The simplest patterns to be matched would be a straight line, a combination of lines, a spot, or a combination of spots. The next higher level of complexity would be a pattern or **template** representing an object such as a printed circuit board, alphanumeric characters, or a layout view of a machine part. The highest level of complexity would be images of the same scene taken from different locations, such as aerial photographs of images taken from different sensors on different satellites.

The matching of patterns involves three types of movement, rectangular, rotational, and warping. Most industrial systems organized in terms of rectangular coordinates match patterns by shifting in the x and y directions and normally do not include the rotation process. The rotation process on a system organized in rectangular coordinates is very complex, as can be seen in Figure 6-11. The shape of the pixels and of features in the image is modified according to the angle of rotation. A vision system can be organized on a polar coordinate data system where polar array sensors are used for the data acquisition. There are no polar array cameras commercially available today. The pixels will be essentially identical for zero and ninety degree directions, but will change as the vector moves away from x and y directions. The polar coordinate system has certain advantages if the need is to scan along a given radius.

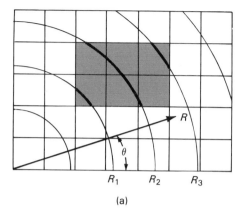

(a)

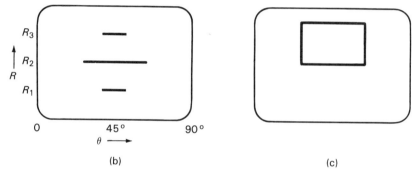

(b)   (c)

Figure 6-11 Image of rectangular object (a) in conventional rectangular coordinates (b) and in polar coordinates (c) displayed on a conventional TV monitor.

The **cross-correlation** between the two patterns can be used as a measure of the degree of match or mismatch. The cross-correlation is determined by calculating the sum of the distance between corresponding pattern points in the two images. The technique is to calculate the cross-correlation factor for one location of the two patterns and then to shift one pattern (called the template) a given amount and recalculate the cross-correlation. The process is repeated over the range of interest, and the location of the best match is found when the sum of the distance is at zero or at a minimum. The sum will be zero as indicated for location D in Figure 6-12 when the pattern and template coincide. The presence of noise or distortion will result in the cross-correlation factor not going to zero, but the minimum value is the best estimate of the optimum location of a match. The matching process can use either binary or multi-gray level data since the magnitude of the intensity at the reference pixel locations is not used. The cross-correlation convolution of the image I and the template T is equivalent to the point-by-point multiplica-

(a)

Summary of results of the template shifting routine for the seven locations.

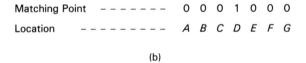

Matching Point	– – – – – – –	0	0	0	1	0	0	0
Location	– – – – – – – – –	A	B	C	D	E	F	G

(b)

Figure 6-12 Pattern matching for a 3 x 1 image: the distance D and the number of matching corresponding reference elements is determined for all locations. The pattern and template are at same height, and movement of template is one unit at a time in a horizontal direction (a). The best fit is indicated to be location D (b) where there is the only matching point condition of all seven locations.

tion of the Fourier transformations of I^* and T and followed by the inverse transformation. This method can be faster for some applications using the FFT algorithims and is found on some systems.

The matching of a 3×1 pattern in the x direction can be achieved by shifting the template from right to left one pixel at a time and calculating the sum of the distances between key registration points. In the example in

Figure 6-12, the pixel containing the *I* is selected as the registration point. The fourth location, case D, provides the minimum normalized cross-correlation factor or distance and has the highest matching point value of any of the locations. The method is extended to a 3 × 3 image pattern in Figure 6-13 where the template is first moved in the *x* direction and then the process is repeated after stepping one pixel in the *y* direction. The output is an expanded matrix containing the number of matching points.

The best match location can be determined by use of either the sum of the distances factor or the maximum number of points in binary images. The cross-correlation or sum of the distances approach provides some insight into the sensitivity of the match to the location in the presence of noise. The best matching point for the template in Figure 6-13 was row 3, column 3. The presence of a noise element in pixel (2,2) of the pattern would produce the results shown in Figure 6-14, Example A. There is one location indicated, but the number of matching points has been reduced. Examples B and C indicate the type of patterns which result in multiple matching points where all the indicated locations result in same degree of accuracy in the match. Example D indicates that the results of matching an outline pattern is one of the best locations. The matching of patterns by rotation in polar coordinates can use the same approach.

The selection of the reference points should be based on the minimum number of unique points which will identify the members of the family of objects which may be in the image.

The examples shown have all been linear, but nonlinear templates can be used.

Industrial machine vision **morphology** is an extension of mathematical morphology whereby set theory methods are used for image analysis. The enhancement of the image features is dependent on their shapes, and pattern recognition techniques are used to determine what features can be measured and how to measure them. The image processing and pattern recognition are incorporated into a single framework. The method is ideally suited for industrial applications where objects are overlapping and a decision must be made regarding the location of the object which is on top. The process makes use of dilation and erosion processing. In the dilation process, a pixel is added to the outer edge of each solid area. Hence the object diameter increases by two pixels, and holes in the object decrease in size. The erosion process is the opposite process. A pixel is removed from the outer edge of each solid area. The object becomes smaller and holes grow larger.

6.5.3 Sequential Edge Detection

The application of vision systems for control purposes requires the ability to identify and track a specific path. The edge detector operators previously discussed processed the image data in a parallel fashion on a global basis for feature identification purposes. The tracking function can be

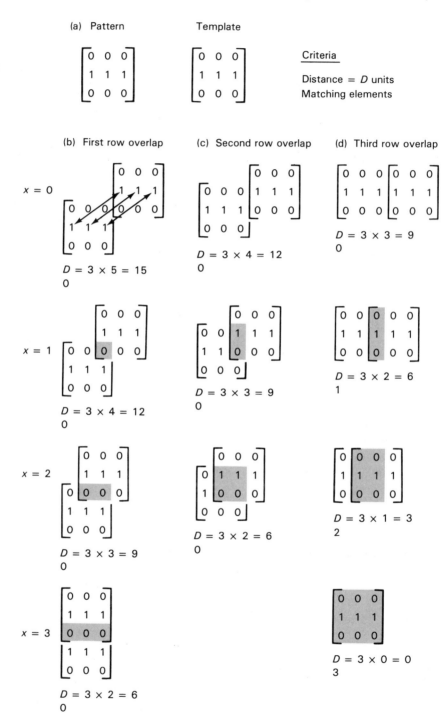

Figure 6-13

(e) Resulting pattern for distance between the three reference
points as x = 0 to 6

Distance *D*   Matching points

First row	15	12	9	6	9	12	15
Second row	12	9	6	3	6	9	12
Third row	9	6	3	0	3	6	9
Fourth row	12	9	6	3	6	9	12
Fifth row	15	12	9	6	9	12	15

$$
\begin{bmatrix}
0 & 0 & 0 & 0 & 0 & 0 & 0 \\
0 & 0 & 0 & 0 & 0 & 0 & 0 \\
0 & 1 & 2 & 3 & 2 & 1 & 0 \\
0 & 0 & 0 & 0 & 0 & 0 & 0 \\
0 & 0 & 0 & 0 & 0 & 0 & 0
\end{bmatrix}
$$

Figure 6-13 Pattern matching for a 3 x 3 image: pattern location is fixed and template is moved horizontally one unit at a time, and the process is repeated for each of the possible vertical rows. The template and pattern (a) are shown various horizontal locations of the one row overlap (b), two row overlap (c), three row overlap (d), and the results summary (e) indicating the best matching position indicated by the lowest *D* and highest mating point location is the third row, fourth position.

(a)
$$
\begin{bmatrix}
0 & 0 & 0 \\
1 & 0 & 1 \\
0 & 0 & 0
\end{bmatrix}
\quad
\begin{bmatrix}
0 & 0 & 0 \\
1 & 1 & 1 \\
0 & 0 & 0
\end{bmatrix}
\quad
\begin{bmatrix}
10 & 8 & 6 & 4 & 6 & 8 & 10 \\
8 & 6 & 4 & 2 & 4 & 6 & 8 \\
6 & 4 & 2 & 0 & 2 & 4 & 6 \\
8 & 6 & 4 & 2 & 4 & 6 & 8 \\
10 & 8 & 6 & 4 & 6 & 8 & 10
\end{bmatrix}
\quad
\begin{bmatrix}
0 & 0 & 0 & 0 & 0 & 0 & 0 \\
0 & 0 & 0 & 0 & 0 & 0 & 0 \\
0 & 1 & 1 & 2 & 1 & 1 & 0 \\
0 & 0 & 0 & 0 & 0 & 0 & 0 \\
0 & 0 & 0 & 0 & 0 & 0 & 0
\end{bmatrix}
$$

Sum of distances   Matching points with
one location

(b)
$$
\begin{bmatrix}
0 & 1 & 0 \\
0 & 1 & 0 \\
0 & 1 & 0
\end{bmatrix}
\quad
\begin{bmatrix}
0 & 0 & 0 \\
1 & 1 & 1 \\
0 & 0 & 0
\end{bmatrix}
\quad
\begin{bmatrix}
1 & 1 & 1 \\
1 & 1 & 1 \\
1 & 1 & 1
\end{bmatrix}
$$

Matching points with
nine locations

(c)
$$
\begin{bmatrix}
0 & 1 & 0 \\
0 & 1 & 0 \\
0 & 1 & 0
\end{bmatrix}
\quad
\begin{bmatrix}
1 & 1 & 1 \\
0 & 0 & 0 \\
1 & 1 & 1
\end{bmatrix}
\quad
\begin{bmatrix}
1 & 1 & 1 \\
1 & 1 & 1 \\
2 & 2 & 2 \\
1 & 1 & 1 \\
1 & 1 & 1
\end{bmatrix}
$$

Matching points with
three locations

(d)
$$
\begin{bmatrix}
1 & 1 & 1 \\
1 & 0 & 1 \\
1 & 1 & 1
\end{bmatrix}
\quad
\begin{bmatrix}
1 & 1 & 1 \\
1 & 0 & 1 \\
1 & 1 & 1
\end{bmatrix}
\quad
\begin{bmatrix}
1 & 2 & 3 & 2 & 1 \\
2 & 2 & 4 & 2 & 2 \\
3 & 4 & 8 & 4 & 3 \\
2 & 2 & 4 & 2 & 2 \\
1 & 2 & 3 & 2 & 1
\end{bmatrix}
$$

Matching points
with one location

Figure 6-14 Complex pattern matching: Effect of a one pixel error in horizontal pattern (a), vertical and horizontal patterns matching (b), subtractive pattern matching (c), and outline pattern matching (d).

accomplished by sequential edge detection where the results of the point selected are contingent upon the results obtained with the operator on the previous point. The application may be to identify and follow the track of the junction of two pieces of metal in a robotic welding operation or for following a given contour in a control process of an industrial manufacturing operation. The metal joint is characterized by two edges with different pixel values for the plates and the space between plates.

The sequential edge detection procedure requires

1. Selection of the starting point,
2. Establishment of a model on the relationship of adjacent points and a procedure for examining adjacent points to determine the path, and
3. Determination of criteria to establish when the path is complete.

The starting point will usually be selected on the basis of prior knowledge of the application; the joining of two plates implies that the track would start and end at the edge of the objects. The tracking of the edge of a hole in an object would indicate the need for a different type of starting point selection criteria; it might be the dimensional coordinates of a pixel.

The model may include provisions for noise in the form of an extra pixel which could direct the tracking process in a wrong direction or for noise in the form of missing pixels which will appear as a break in the edge. The path can be considered as having four directions from any point. They are the direction of the previous point and the other three new directions. The task is to determine which direction has the highest probability of being a continuation of the track. Again, weighing factors may be developed based on the physical constraints of the task, such as the path has no change in direction of more than 30 degrees over a three inch distance.

REFERENCES

1. James C. Solinsky, "Correlation or Morphology: It Depends on the Model Vision," *Vision Technology,* October, 1985.
2. Stanley N. Lapidus, "Gray-Scale and Jumping Spiders," *Vision Technology,* January, 1985.
3. Stanley R. Sternberg, "Industrial Morphology," *Interobot West,* October, 1984.
4. J. Serra, *Image Analysis and Mathematical Morphology,* New York: Academic Press, 1982.
5. H. C. Andrews and K. Casparie, *Computer Techniques in Image Processing,* New York: Academic Press, 1970.
6. Edward R. Dougherty and Charles R. Giardina, *Matrix Structured Image Processing,* Englewood Cliffs: Prentice Hall, 1987.

EXERCISES

1. Apply a Robert's gradient Operator to the input matrix and determine the output matrix in the area.

$$\begin{bmatrix} 2 & 4 & 4 & 5 & 6 \\ 1 & 5 & 7 & 13 & 2 \\ 2 & 11 & 13 & 2 & 14 \\ 3 & 3 & 12 & 11 & 1 \\ 14 & 12 & 2 & 3 & 4 \end{bmatrix}$$

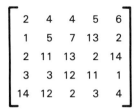

Input

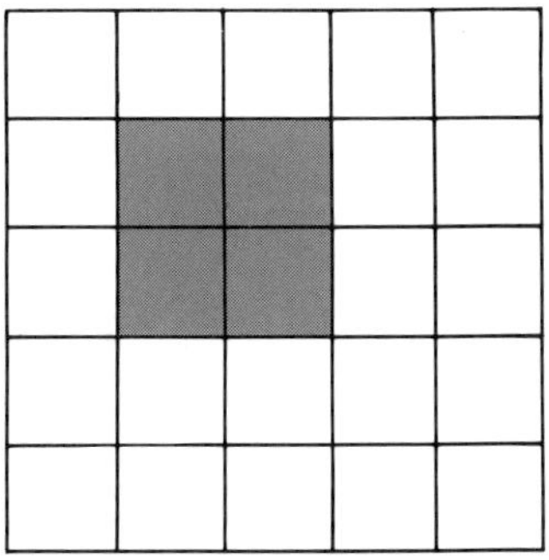

Output

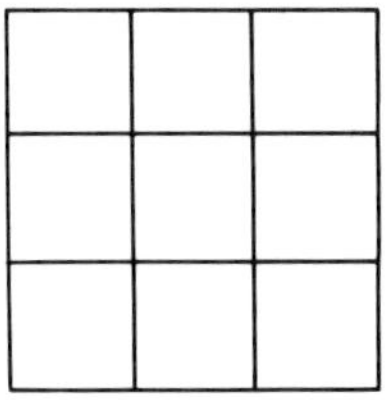

High pass filter

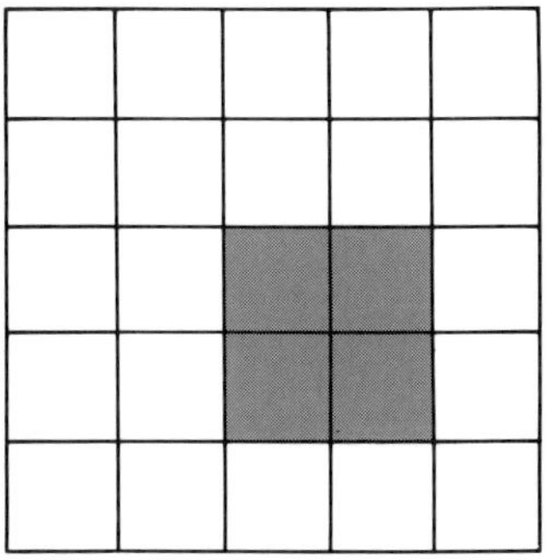

Output

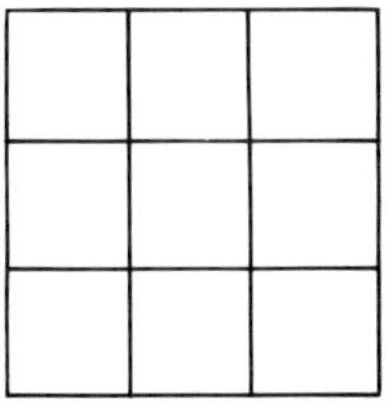

Low pass filter

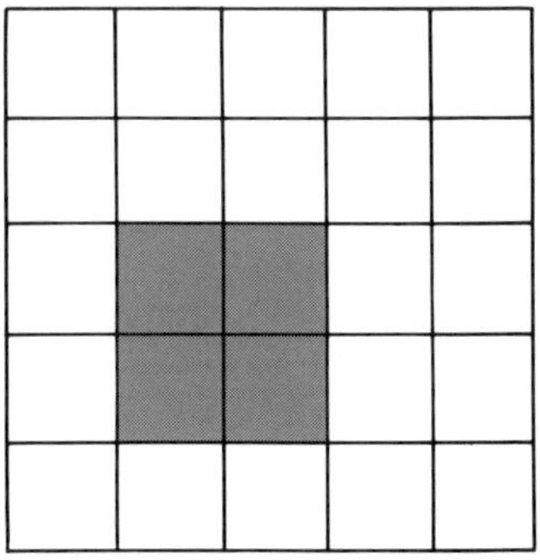

Output

2. Define a high pass filter and determine the output in the areas shown for the input matrix of 1.
3. Define a Low pass filter and determine the output in the areas shown for the input matrix of 1.
4. Write a program which will apply either a low pass or a high pass filter to the data in a given input image. The program should produce a histogram of the image before and after the filtering process.

Standard Input Matrices Data

(a)
$$\begin{bmatrix} 2 & 2 & 2 & 2 & 2 \\ 2 & 2 & 2 & 2 & 2 \\ 2 & 2 & 2 & 2 & 2 \\ 4 & 4 & 4 & 4 & 4 \\ 4 & 4 & 4 & 4 & 4 \end{bmatrix}$$

(b)
$$\begin{bmatrix} 2 & 2 & 5 & 6 & 2 \\ 2 & 2 & 5 & 6 & 2 \\ 2 & 0 & 4 & 6 & 2 \\ 4 & 4 & 4 & 8 & 4 \\ 4 & 4 & 4 & 5 & 4 \end{bmatrix}$$

(c)
$$\begin{bmatrix} 8 & 8 & 8 & 8 & 9 & 10 \\ 8 & 0 & 8 & 9 & 10 & 10 \\ 8 & 8 & 9 & 10 & 10 & 9 \\ 8 & 9 & 10 & 10 & 9 & 9 \\ 9 & 10 & 10 & 9 & 9 & 9 \\ 10 & 10 & 9 & 9 & 9 & 9 \end{bmatrix}$$

(d)
$$\begin{bmatrix} 11 & 11 & 12 & 13 & 13 & 13 & 10 & 10 \\ 11 & 11 & 12 & 13 & 13 & 13 & 10 & 10 \\ 11 & 11 & 12 & 13 & 13 & 13 & 10 & 10 \\ 11 & 11 & 12 & 13 & 13 & 13 & 10 & 10 \\ 12 & 12 & 13 & 14 & 14 & 14 & 11 & 11 \\ 13 & 13 & 14 & 15 & 15 & 15 & 12 & 12 \\ 13 & 13 & 14 & 15 & 15 & 15 & 12 & 12 \\ 12 & 12 & 13 & 14 & 14 & 14 & 11 & 11 \end{bmatrix}$$

(e)
$$\begin{bmatrix} 8 & 6 & 4 & 4 & 4 & 4 & 6 & 8 & 8 & 8 & 6 & 4 \\ 8 & 6 & 4 & 4 & 4 & 4 & 6 & 8 & 8 & 8 & 6 & 4 \\ 8 & 6 & 4 & 4 & 4 & 4 & 6 & 8 & 8 & 8 & 6 & 4 \\ 8 & 6 & 4 & 4 & 3 & 5 & 7 & 8 & 8 & 8 & 6 & 4 \\ 8 & 6 & 4 & 4 & 4 & 6 & 8 & 9 & 8 & 8 & 6 & 4 \\ 6 & 4 & 3 & 3 & 5 & 7 & 7 & 7 & 7 & 7 & 7 & 7 \\ 4 & 2 & 2 & 4 & 6 & 8 & 8 & 8 & 8 & 8 & 8 & 8 \\ 1 & 2 & 3 & 5 & 7 & 8 & 9 & 9 & 9 & 9 & 9 & 9 \\ 2 & 3 & 4 & 6 & 8 & 9 & 10 & 10 & 10 & 10 & 10 & 10 \\ 3 & 3 & 5 & 7 & 7 & 7 & 7 & 8 & 9 & 9 & 10 & 11 \\ 4 & 4 & 6 & 8 & 8 & 8 & 8 & 9 & 10 & 12 & 13 & 14 \\ 3 & 5 & 7 & 7 & 7 & 7 & 8 & 10 & 12 & 14 & 14 & 15 \end{bmatrix}$$

5. Design a high pass digital filter and apply it to the input matrix c; determine the effect of the filter.

6. Design Robert's, Laplacian, and Sobel filters and apply them to the data in input matrices a, b, and c.

7. Apply the Robert's filter to the data obtained from the vision system. Discuss the before and after values.

8. Design a template which will identify the letters I, B, T, and O.

9. Write a program which will identify the four letters in Problem 8, indicate which letter is contained in the image, and print the letter and locations of the match.

10. Design a mask which will erode an area.

11. Design a mask which will dilate an area.

12. Design a program which will count the pixels between two edges.

13. Design a program which will count the pixels in an area.

14. Design a program which will count the number of holes in an object.

15. Plot the histogram of an image before and after using a high pass digital filter on the images d and e to eliminate some of the bias.

16. Match the columns.

1. Low pass filter	_____	a. Consists of a positive and a negative edge.
2. High pass filter	_____	b. Process which can be used to identify object in image.
3. Laplacian operator	_____	c. Can be used to eliminate noise effects: characterized by coefficients in matrix being positive and summing to 1.
4. Sobel edge detector	_____	d. Determines the new pixel value based on the value of pixels in a 2×2 region.
5. Sequential edge detection	_____	e. Can be used to eliminate effect of relatively constant background lighting: characterized by coefficients in matrix being positive, zero, or negative, and summing to 0.
6. Robert's gradient	_____	f. High pass filter based on the slope, or second partial derivative, of image intensity in x and y directions.
7. Line		g. Operator which utilizes gradient of pixel values as basis for edge identification; does not utilize value at the point in the process.
8. Pattern matching	_____	h. Selects next point in track based on the results of the operator on previous points.

7

ACQUIRING
THE INFORMATION:
BAR CODING

7.0 INTRODUCTION

The automatic acquisition of information on an object is critical to the task of using machine vision to improve productivity. The task is greatly simplified if the object is coded by an artificial identifier rather than having to look at the natural features of the object and use one or more of them for the identification parameters. Bar coding is a form of artificial identifier. It is a machine readable code consisting of a pattern of black and white bars and spaces in defined ratios which represent alphanumeric characters. A sensor scans the bar code symbol and converts the visual image into an electrical signal. The optics involved may be a simple lens or a **fiber-optic** transmission system which transmits the light signal to a detector located in a unit some distance away from the measurement point. The information encoded in the electrical signal of the bar code is then processed by a decoder which is programmed to obtain the desired information in a way similar to the way information from your eye is processed by your brain.

Bar coding is the easiest, the most cost-effective and the most reliable method of identifying and entering information into a computer based information system. Today, bar coding is the accepted method of acquiring tracking information on products both on the manufacturing line and in the distribution system.

The initial efforts in bar code technology on a relatively large scale occurred in the 1960s with major efforts to automate the postal system and to improve the railroad car tracking system. The railroads were interested in replacing the individual who would stand beside the track and record the number on each passing freight car. The initial railroad effort was aimed at developing a color bar code as the technology was still in its infancy and neither the sensors nor portable computing power were available for the system we have today.

The post office effort involved automatic sorting and routing of letters. Much effort was concentrated on automatic recognition of handwritten characters and the bilevel or bar and half-bar patterns that are still being used by the U.S. Post Office for machine vision sorting operations. The banking and credit card industry went to a similar technique where machine readable characters were imprinted on the document at an early part of the processing cycle. The need of the military to track or trace components in complex missile systems was met using a human readable alphanumeric identification system on all the critical components.

The interleaved 2 of 5 bar code was formulated in 1972. The grocery industry adopted the Universal Product Code (UPC) in 1973 as a standard for item identification. The code 39 symbology was invented and placed in public domain in 1974. This was followed by Code 11 in 1977 and Code 93 in 1982. One of the major turning points in the development and application of bar coding was acceptance of 3 of 9 or Code 39 by the Department of Defense in 1981. Code 39 was adopted in 1984 by the Health Industry Bar Code Council (HIBCC), by the Automotive Industry Action Group (AIAG), and by the American Paper Institute (API). The AIAG also adopted the interleaved 2 of 5 in the same year.

The bar code sensor can either be hand held or machine mounted. With the hand held method, the operator requires only minimal training and can enter information at a speed far greater than a trained typist and with a much higher degree of accuracy. This method also provides great flexibility concerning the location of the bar code symbol on the product. A minimum error rate on the order of one character in three million is achievable with proper symbology, good labels, and reasonable equipment. This contrasts to an error rate of one error in three hundred characters entered manually from a keyboard and one error in ten thousand for optical character recognition systems. The error rate can be further reduced for machine vision systems by using check digits and other data security techniques if the application warrants the additional cost involved.

A number of different bar code symbologies are used in industry today and will continue to be used, since each has special characteristics required by different sectors of the economy. The reader or scanner contains an illumination source and a detector to sense the illumination reflected from the surface containing the bar code label. In addition, reader systems must have decoding capabilities to produce the appropriate output signal according to the bar code symbology being used. The output is displayed on a **light**

TABLE 7-1   Bar Code Readers

Type	Description	Cost of Unit
1. LED	Red light emitting diode	Lowest
2. IR	Infrared	Medium
3. Narrow Band	Laser	Highest
4. Fiberoptic	Environment	High

emitting diode (LED) display or input to a computerized system using a standard RS 232link.

Laser moving-beam scanners introduced in the early 70s, coupled with the widespread application of microprocessors, led to the development of autodiscriminating capabilities, the ability to read different bar code symbologies interchangably. The autodiscriminating ability permits greater flexibility by allowing use of more than one symbology in a product line or industry; for example, the drug industry uses UPC for certain functions and Code 39 for new functions.

7.1 TYPES OF READERS

Four basic types of bar code readers are listed in Table 7-1. Each reader has special cost and operating characteristics.

7.1.1 Light Emitting Diode (LED)

The LED reader, described in Figure 7-1, is the lowest in cost, but is affected by ambient lighting conditions found in the work place. It must be essentially in contact with the surface and is subject to degradation if the bar code is not clean. The depth of field is on the order 0.075 inch.

7.1.2 Infrared (IR)

The infrared bar code reader is less affected by ambient lighting conditions in the workplace and provides greater immunity to dirt and contamination than the LED units.

7.1.3 Coherent Light (Laser)

The laser units are currently the most expensive of the readers, but they provide the greatest depth of field. This is on the order of three inches. Hence, the measurements can be made at a distance. The 750 nanometer laser wavelength of the unit is not visible, and the units contain a red LED source to help the operator aim the scanner. The unit contains the elements to move the beam back and forth to scan the bar code.

Welch Allyn

WANDS Series Model SRD
Digital Red LED Scanner

DESCRIPTION

The Welch Allyn WANDS Series Model SRD is a rugged, light weight, hand held digital bar code scanner. The visible red illumination and high quality electronics are suitable for numerous bar code reading applications.

APPLICATIONS

The SRD is an optical and electronic interface between a label and a bar code reading terminal. The versatility of the SRD makes it suitable for heavy industrial use as well as office and retail applications.

FEATURES

- **EXCELLENT DEPTH OF FIELD**
  The 0.65'' (1.6mm) depth of field permits scanning in contact, from 0° to 45°. The SRD optics also permit scanning through heavy laminates.

- **STAINLESS STEEL TIP**
  The smoothly contoured tip glides easily across bar code labels without damage and a window protects the scanner optics from dust and other contaminates. The tip is easily changed in the field without tools.

- **ELECTRONICS**
  A single voltage circuitry with external enable line operates over a wide input voltage range. A choice between open collector and TTL output signals is available.

- **OPTIONAL SWITCH**
  An optional push-to-operate switch is available. The finger-operated switch is active at all positions around the tube circumference.

- **HIGH SCAN SPEED**
  The optics and electronics are designed to read high density bar codes at speeds of 2 to 50 inches per second (5 to 130cm/sec).

- **VISIBLE RED LIGHT SOURCE**
  The visible red LED operates in the 660nm range, permitting the scanner to see colored inks which are invisible to infrared light sources.

- **LIGHT WEIGHT**
  The combination of light weight construction and highly flexible cord prevents operator hand fatigue during extended periods of use.

- **CHOICE OF READING APERTURE**
  The SRD is available with a choice of high, medium or low resolution optics to meet needs imposed by various printing methods and print quality.

SPECIFICATIONS

PRINCIPAL FOCAL POINT

Factory set at 0.04'' (1mm), although the user will notice no change of operation on either side of the principal focal point within the specified depth of field.

Preferred Orientation — Tip in contact with target and projection of scanner axis parallel to bar code lines. Roll angle: $\theta_R = 20°$. The 45° usable tilt angle is reduced to 35° on high density codes.

DEPTH OF FIELD — 0.065'' (1.6mm)
Depth of field is taken along the optical axis.

OUTPUT CHARACTERISTICS
TTL compatible or open collector, black high.

SUPPLY VOLTAGE
4.3 to 6.0V, less than 15mV ripple and noise.

POWER CONSUMPTION
Nominal Power Requirements:
Enable Low 15mA
Enable High 35mA
With Switch Option:
On 35mA
Off 0mA

PHYSICAL CHARACTERISTICS
7.6 inch (19.3cm) from the tip to the end of the strain relief with a 0.5 inch (1.3cm) diameter. Body and tip are stainless steel.

ENVIRONMENTAL
Operation: -4°F to 104°F (-20°C to 40°C)
Storage: -40°F to 158°F (-40°C to 70°C)
Humidity: 95% maximum non-condensing

ELECTRICAL CONNECTIONS
Red Wire — +5VDC
White Wire — Common
Braid — Shield (EMI/RFI)
Blue Wire — Enable in
Black Wire — Digital out

CABLES & CONNECTIONS
Standard: 6' (1.8m) coil (extended) or straight cord. Special terminations or lengths up to 20' (6m) available by quotation. Shielded/Coiled cord with connector recommended.

OPERATING MODES
(For Non-Switch Operation)

MODE	+V	ENABLE	DIGITAL OUTPUT
LEDs off	H	L	H
Active	H	H	H/L
* Not Defined			

NOTES

1. When +VDC is applied, typically via a saturated transistor, there is a 4 millisecond maximum turn-on time; if the Enable is low, the output is not defined.

2. The Enable input must be pulled below 0.5 volts at 1mA and above 3.9 volts at 1mA. Enable current will be nominally 1.5mA. A recommended driver is the CD4049. When taken high, the LED turns on and within 0.5 milliseconds the scanner output assumes the correct state appropriate for what is being viewed.

Figure 7-1 LED reader specifications.

7.2 BAR CODE READERS

The bar code reader is one of the simplest vision system input devices. The reader is a one-dimension off-on sensing device. The unit is in the low, or off, state when the field of view is black and is in the high state, or on, when the field of view is white (Figure 7-2).

The reader or scanner serves as the eye of the bar code system by converting a visual black and white image into an electrical signal. Some of the considerations when selecting a reader are

1. Analog or digital output
2. Type of illumination
3. Contact or noncontact
4. Fixed or moving beam
5. Environment

The proper selection of the reader is dependent upon its compatability with the application, the label, the decoder, and the total system. The decision on analog or digital output is determined by the input of the decoder part of the system. The type of illumination is dependent on the environment, the type of application, and the type of labels. Infrared is more tolerant of dirt and contamination frequently found in the manufacturing environment. The type of ink used in the label must be coordinated with the type of illumination used. The choice is fixed or moving beam and contact or noncontact is determined by the application. Contact units offer lower initial costs but are subject to wear and, hence, maintenance is more costly. Noncontact devices offer a greater depth of field and provide more flexibility with regard to mounting.

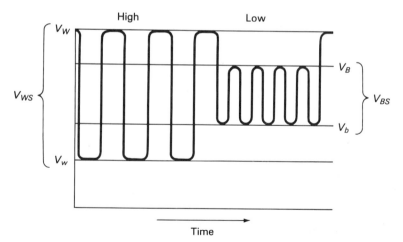

Figure 7-2   Bar code reader output voltage: $V(ws)$ is voltage output when viewing white space and $V(bs)$ is voltage output when viewing a bar or black space.

The bar code must be completely scanned as the clear zones and the start and stop patterns are used in the decoding process for control purposes. A relatively smooth scan at constant speed is important. As a practical matter, it is generally difficult to obtain much acceleration in the short span lengths.

Associated with the bar code reader is a unit which decodes the electrical signal generated by the reader. It uses the timing relationship of the electrical signals generated by the bars and spaces to decode the encoded information in the bar code symbol. Some of the factors which should be considered in selecting a decoder are

1. Scanner compatibility
2. Output required
3. Portability
4. Need for display or printout
5. Need for key entries
6. Number and variety of codes to be read
7. Environmental factors

Bar code reading and decoding equipment have varying degrees of technical sophistication. Units may have automatic gain control to compensate for poorly printed or faded labels, autodiscrimination of multiple codes, and system interface features for multiple station application.

7.2.1 Aperture

All scanners have an aperture; this is the opening through which the light reflected from the surface passes to the detector element. The aperture must be smaller than the width of the bars being read but must be as large as possible, consistent with the sensitivity of the detector element.

7.2.2 Aperture Effect

The aperture of the unit is related to the width of the bars to be read. The variation of output as a function of the aperture is given in Table 7-2.

7.2.3 Resolution Index

The resolution index (RI) of optical bar code scanners is defined as a ratio of the *narrow bar signal* to the *wide bar signal* multiplied by 100 percent.

The narrow bar signal (VD) is defined as the magnitude of the electrical signal which is generated while scanning the narrowest bars of a test target. The wide bar signal (VS) is defined as the difference between the magnitudes of the electrical signal which is generated on the ''pure'' black area of a target and the electrical signal generated as the ''pure'' white area of a target. This is

TABLE 7-2   Effect of Aperture Size

Parameter	Minimum	Typical
Wide Bar Signal	Volts	Volts
0.004 inch Aperture	100	150
0.006 inch Aperture	200	300
0.008 inch Aperture	400	600
0.010 inch Aperture	620	930
0.012 inch Aperture	900	1350
Narrow Bar Signal		
0.004 inch Aperture	50	90
0.006 inch Aperture	100	180
0.008 inch Aperture	200	360
0.010 inch Aperture	310	558
0.012 inch Aperture	450	810

equivalent to the signal produced by infinitely wide bars and spaces. This can be expressed as the equation

$$RI = VD/VS \times 100\%$$

where

$$VD = Vw - Vb$$
$$VS = VW - Vw$$

Measurements are referenced to the National Bureau of Standards NBS 1010A test target having a bar width 6.001 inches larger than the scanner aperture; to make the units independent of the normal scanner velocity, the "rise time" and the "fall time" to go between 10 percent and 90 percent of the signal is on the order of 40 microseconds.

7.3 BAR CODE SYMBOLOGY

A number of different symbologies have been developed and are in use today. Their characteristics vary in terms of error checking characteristics, density or characters per inch, the type of characters (such as alpha or alphanumeric) which can be encoded, and the industry application. Five principal bar codes in common use today are

> Code 3 of 9, (Code 39)
> Interleaved 2 of 5
> Codabar
> UPC (Universal Product Code)
> EAN (European Article Numbering)

In some cases, a type of code is used because of special performance characteristics; in other cases it is adopted by an industry based on historical or special commercial interests and is so widespread that there is little likelihood that it will be replaced by some other code.

Uniform Symbology Specifications (USS) were developed by the Automatic Identification Manufacturers, Inc., (AIM) Pittsburgh, Pa., as a guide to aid the manufacturer, the consumer, and the general public. The uniform symbology specifications may be revised or withdrawn at anytime by AIM and users should check with them concerning the latest specifications.

7.3.1 Code 3 of 9

The 3 of 9 code has been adopted for use by the U.S. Department of Defense, the Post Office, the Health Industries Bar Code Council, and many industrial users. This should have the widest future application since DOD is a very large purchaser in the United States and can require use of the code on the products it uses.

The label contains three wide elements (bars or spaces) out of a total of nine elements (Figure 7-3). Five of the elements are dark bars and four are light bars. Three elements are 2 units wide and six are 1 unit wide, for a total of 12 units per character.

Code 39 is a bar code symbology with a full alphanumeric character set, a unique start and stop character, and seven special characters.

The symbol will contain

1. One quiet zone at each end
2. Start and stop characters
3. Data string of one or more characters

The highest printing density is on the order 9.8 alphanumeric characters per inch and is highly immune to substitution errors. It can be used to encode an upper-case character set as well as other special characters. The length is variable.

A human readable interpretation of the data characters is generally

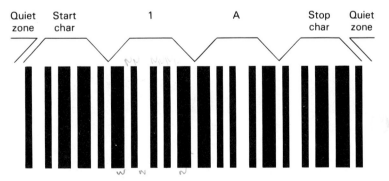

Figure 7-3   Encoding the message "1A."

Figure 7-4   Example of human readable bar code.

printed along with the USS-39 symbol encoding them. Character size and font are not specified, and the message may be printed anywhere in the area surrounding the symbol, as long as quiet zone boundaries are not violated.

Start and stop human readable interpretation (HRI) characters should not be printed. Examples of human readable bar code labels are given in Figure 7-4.

Construction.   The width of bars and spaces in the 3 of 9 system are in binary form. The narrow bars and spaces represent a zero and the wide bars and spaces represent a one. Since three of the nine elements represented are wide and six are narrow, each bar code symbol in the data field will contain three 1s and six 0s.

The bar code symbol will contain a start code at the left end and a stop at the right end of the entire character string in the label or message. The region between the start and stop is the data field and can contain up to 32 alphanumeric, as shown in Figure 7-5. The number of characters is limited only by equipment capability.

/Start---/---Data   Field---/---Error   Checking

Character---/---Stop/

(32 Characters)   (1 Character-Optional)

Figure 7-5   Layout of 3 of 9 data string.

Bar code character.   A sample bar code data field for characters 1 and 4 would be a combination of the bars and spaces as shown in Figure 7-6. The complete 3 of 9 character set is given in Table 7-3.

Each bar or space can be one of two alternative widths, referred to as wide and narrow. The particular pattern of wide and narrow bars determines the character being coded. In all cases, each character consists of three wide and six narrow elements.

Table 1 defines all of the USS-39 character assignments. In the table columns headed Bars and Spaces, the character 1 is used to represent a wide element, and a 0 to represent a narrow element. Each character is separated by an intercharacter gap as illustrated in Figure 7-7.

7.3.2 Interleaved 2 of 5

Only numeric data can be encoded with the interleaved 2 of 5 code. The highest density is 18 characters per inch. The code is prone to error unless special conditions are imposed on the numbers being encoded. Pairs of characters are encoded in the symbol. The first character of the pair is represented by the bars and the second character is represented by the spaces. The Interleaved 2 of 5 symbology is used by the United States food

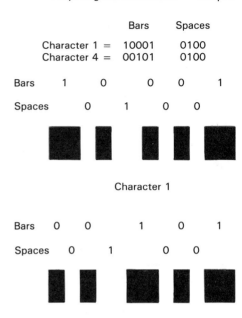

	Bars	Spaces
Character 1 =	10001	0100
Character 4 =	00101	0100

Bars   1   0   0   0   1

Spaces   0   1   0   0

Character 1

Bars   0   0   1   0   1

Spaces   0   1   0   0

Character 4

Figure 7-6 Layout of 3 of 9 bar code characters.

TABLE 7-3   USS-39 Character Structure

CHAR.	PATTERN	BARS	SPACES	CHAR.	PATTERN	BARS	SPACES
1		10001	0100	M		11000	0001
2		01001	0100	N		00101	0001
3		11000	0100	O		10100	0001
4		00101	0100	P		01100	0001
5		10100	0100	Q		00011	0001
6		01100	0100	R		10010	0001
7		00011	0100	S		01010	0001
8		10010	0100	T		00110	0001
9		01010	0100	U		10001	1000
0		00110	0100	V		01001	1000
A		10001	0010	W		11000	1000
B		01001	0010	X		00101	1000
C		11000	0010	Y		10100	1000
D		00101	0010	Z		01100	1000
E		10100	0010	,		00011	1000
F		01100	0010	.		10010	1000
G		00011	0010	SPACE		01010	1000
H		10010	0010	@		00110	1000
I		01010	0010	$		00000	1110
J		00110	0010	/		00000	1101
K		10001	0001	+		00000	1011
L		01001	0001	%		00000	0111

*Denotes the special start/stop code character

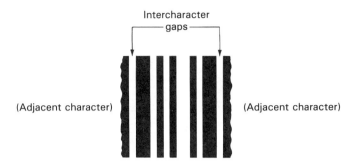

Figure 7-7   USS-39 A character.

and drug industry for shipping containers as well as by the auto and paper products industries. The characteristics of the Interleaved 2 of 5 (or USS-I 2/5) is given in Table 7-4.

Pattern description.   A symbol for a number consists of bars and spaces for each character enclosed by special start and stop patterns and quiet zones. The character code is based on the format that decimal digits are represented by five binary characters (four weighted bits and a parity bit) in which only two of the five bits are one. Therefore, only two of the five bits are wide. The four-element pattern on the left and the three-element pattern on the right end of the number permit bidirectional decoding of the symbol.

A number encoded must always contain an even number of characters. If the number to be encoded has an odd number of digits, a zero is added to the leading edge to produce the even number of digits. The number to be encoded must first be grouped into pairs of adjacent characters as shown in Figures 7-8 and 7-9.

The bar code symbol frequently will include human readable characters adjacent to the bar codes marks to facilitate checking when no bar code reader is available. The numeric characters are printed above or below the bar code. The start and stop characters are not printed. The length of the bar code symbol can be calculated by

$$L = (P(4N + 6) + 6 + N)X + 2Q$$

where P is the number of character pairs, N is the wide to narrow element ratio, X is the width of a narrow element, and Q is the width of the quiet zone.

**TABLE 7-4   Characteristics of Interleaved 2 of 5
or USS-I 2/5 Bar Codes**

Type of Characters: Numeric
Length: Variable, but must be even number of digits
Decoding: Bidirectional
Maximum data density: 18 characters/inch
Special characters: Different start and stop Patterns

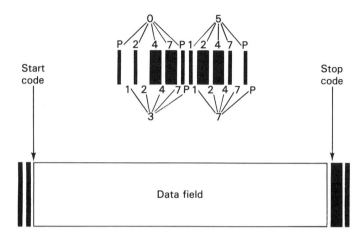

Figure 7-8 Example of Interleaved 2 of 5 bar code.

Data Element.   The pairs of characters illustrated in Figure 7-10 are 5,5 in Example A and 1,4 in Example B.

Special pattern requirements.   The leading and trailing ends of the symbol are identified by special start and stop patterns as illustrated in Figure 7-11. The start pattern is located on the left end of the data field and consists of four narrow elements, alternating narrow bars and spaces. The stop pattern is located on the right end of the data field and consists of wide bar and two narrow elements. In addition, there are two quiet zones adjacent to the start and stop patterns as shown in Figure 7-12. The total character set for 2 of 5 bar code symbol is given in Figure 7-13. Wide bars and spaces are considered as binary 1 and narrow bars and spaces are binary 0. Each data character is given by five binary elements: two of the five are binary 1s.

7.3.3 Codabar

The codabar system can be used to encode numeric data, the six special characters $, -, :, /, ., and +, and four start-stop characters, A, B, C, and D. The codabar symbology provides for variable length data strings. The key characteristics are given in Table 7-5. Each data element contains two or three wide elements out of seven as shown in Figures 7-14 and 7-15.

Each codabar label contains two quiet zones, a start and a stop character, and a variable length data field up to 32 characters as illustrated by Figures 7-15 and 7-16.

Each character is represented by a group of seven units: four bars and three interbar spaces. The ones are represented by bars and spaces two elements wide and zeros, by bars and spaces one element wide.

The four stop-start characters can be used to encode different types of information. The symbol contains different combinations of wide bars and/or

Figure 7-9 Interleaved 2 of 5 bar codes.

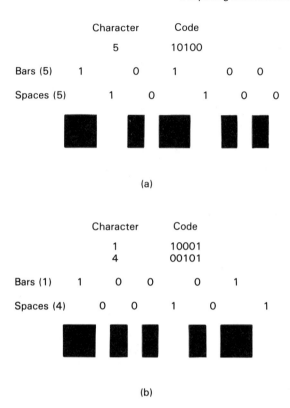

Figure 7-10   Layout of Interleaved 2 of 5 bar code elements.

wide spaces, according to the character. The quiet zone between symbols is ten times the width of the narrowest bar; hence, it is easy to identify characters.

The four start-stop characters A, B, C, and D are stand alone entities and can be used as either start or stop characters; the same character can be used for both functions. This permits the development of additional information elements by using different combinations of the start-stop elements. The characters within the symbol are separated by an intercharacter gap consisting of one narrow space element. A typical character data string is illustrated by Figure 7-17. The element pattern for all the codabar characters is given in

Figure 7-11   Start and stop pattern in 2 of 5 bar code data string.

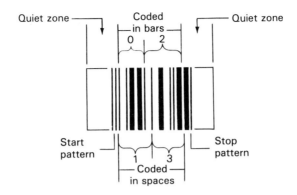

Figure 7-12   Complete 2 of 5 bar code for number 0123.

Figure 7-18, where the wide elements are considered the equivalent of a binary 1 and the narrow elements are considered the equivalent of a binary 0. Each character can be represented by a unique binary 7-bit set shown in the Figure 7-18.

The highest standard print density is 12.8 characters per inch and is essentially immune to substitution errors. It finds use in applications such as labeling of blood samples and where the avoidance of errors is important. It is used in the photo processing industry, automated library circulation systems, and parcel tracking.

(a)

2 of 5 Bar Code Character Set

Number (Decimal)	2 Of 5 code (Modified BCD)	Binary Value
0	00110*	6
1	10001	17
2	01001	9
3	11001	25
4	00101	5
5	10100	20
6	01100	12
7	00011	3
8	10010	18
9	01010	10
Start Character	00	0
Stop Character	10	2

* Exception to modified BCD Code

(b)   Layout of Character String

Start-----/-----Data Field-----/-----Stop

(Variable length)

Figure 7-13   Character set for Interleaved 2 of 5 bar code (a) and layout of data string (b).

TABLE 7-5   Codabar Characteristics

Characters are the ten digits: 0 to 9,
    the six special characters $, -, :, /, ., and +,
    and the four start-stop characters A, B, C,
    and D.
Variable length data set
Bidirectional decoding
Maximum data density—12.8 characters/inch

7.3.4 Universal Product Code (UPC)

The UPC Bar Code character set contains numerics and three special start, stop, and center characters. Each character is made up of four elements, two bars and two spaces covering a total of seven modules. There are two basic data string formats as shown in Figure 7-19.

The Universal Product Code is used to encode item identification and manufacturer numbers on consumer products in the United States. Its use is governed by the UPC council and is not in the public domain.

The number system character indicates the specific and special use of the data. The assignments of number systems characters is given in Table 7-6.

The UPC character set is given in Table 7-7. Sample data string for character 5 using the type 2 format and regular UPC Code from Table 7-7 is given in Figure 7-20. Bars represent ones and spaces represent zeros in the bar code symbol.

There are several modifications such as UPC-A, found on grocery items, and UPC-E zero suppressed format. It offers low data error security and is a fixed length code. Other formats of UPC have been developed for the credit card applications. UPC finds wide use where simple identification is needed but effect of an error is negligible.

TABLE 7-6   Number System Characters

Character	Use
0	Regular UPC code
2	Random weight items such as produce and meats
3	National health-related items code and national drug code
4	In-store working of nonfood items with check code digit protection; also for uses without code format restrictions
5	Coupons
All Others	Future use

Figure 7-14   Codabar.

7.3.5 European Article Numbering (EAN)

The European Article Numbering code is used in Europe where it is the counterpart of UPC. The code is used for numeric data and is shown in Figure 7-21 in conjunction with the UPC code.

There are a number of modifications such as EAN-13 which contains data on the country of origin as well as the item identification number. The details on the above five widely used codes will provide insight on the different characteristics and information it is possible to include in a bar code label.

7.4 BAR CODE LABEL GENERATION

It is common practice in many industries to generate the bar code labels at the site using a computer printer. Regardless of the code, it is important that the bar code label be printed to standard specifications to minimize the errors by users. The bar code label generation can be monitored with special quality

TABLE 7-7   UPC/EAN Character Set

Character	Left Field Characters Odd Parity	Right Field Characters Even Parity
0	0001101	1110010
1	0011001	1100110
2	0010011	1101100
3	0111101	1000010
4	0100011	1011100
5	0110001	1001110
6	0101111	1010000
7	0111011	1000100
8	0110111	1001000
9	0001011	1110100

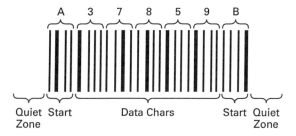

Figure 7-15   Codabar symbol encoding "A37859B."

assurance code bar measurement instruments such as the CODA, SCAN/
AUTO, and SCAN systems, manufactured by RJS Enterprises, a Signode
Corporation Company. The system checks the following four key parame-
ters: layout, encodation, print contrast, and bar and space widths.

There is a normal variation in bar and space width on a label due to the
printing process.

7.4.1 Bar Code Label Production

Bar code labels can either be purchased preprinted, or they can be
printed on site. The method of production of the bar code label will be
dependent upon its intended use.

Preprinted bar code labels can be purchased from suppliers whose
equipment ranges from state-of-the-art photographic-type to offset printing to
standard dot matrix printing.

Preprinted labels.   Preprinted labels can be purchased but should be
subject to quality control. Advantages to using preprinted labels are

- High density bar code labels are available,
- Precise tolerances can be maintained,
- Labels can be laminated or otherwise protected,
- Labels can be produced on non-paper substrates,
- There is no capital investment required for printing hardware, and
- A wide choice of adhesives and packaging is available.

The disadvantages associated with preprinted labels are

- Higher unit cost
- Necessity to predetermine label content

Preprinted labels tend to be of higher quality than on-site label generation product and, in general, use up less of the total tolerance of a bar code
system.

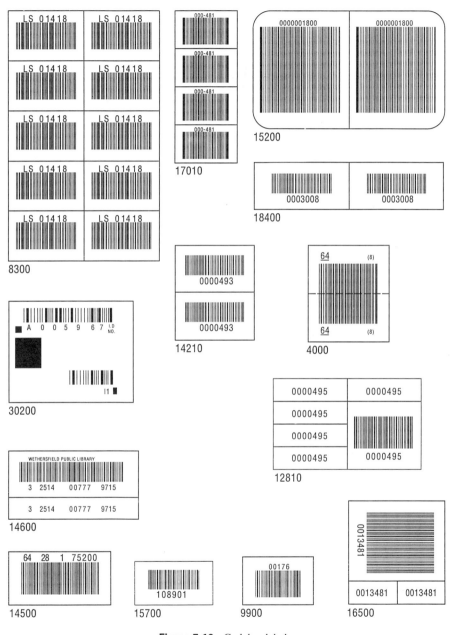

Figure 7-16 Codabar labels.

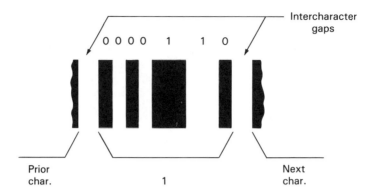

Figure 7-17 Codabar character data string.

On-site label generation.   Bar code labels can be produced on-site using one of several different techniques. The quality is generally lower than that of preprinted labels. However, production flexibility is increased.

THERMAL PRINTERS
- Paper cost is relatively high.
- Medium to low density labels can be printed.
- Moderate label durability is typical.
- Equipment cost is low to medium.

Encoded Character	Binary Representation (B S B S B S B)	Bar and Space Pattern
0	0 0 0 0 0 1 1	
1	0 0 0 0 1 1 0	
2	0 0 0 1 0 0 1	
3	1 1 0 0 0 0 0	
4	0 0 1 0 0 1 0	
5	1 0 0 0 0 1 0	
6	0 1 0 0 0 0 1	
7	0 1 0 0 1 0 0	
8	0 1 1 0 0 0 0	
9	1 0 0 1 0 0 0	
,	0 0 0 1 1 0 0	
$	0 0 1 1 0 0 0	
:	1 0 0 0 1 0 1	
/	1 0 1 0 0 0 1	
.	1 0 1 0 1 0 0	
+	0 0 1 0 1 0 1	
A	0 0 1 1 0 1 0	
B	0 1 0 1 0 0 1	
C	0 0 0 1 0 1 1	
D	0 0 0 1 1 1 0	

Figure 7-18 Codabar character set and binary representation.

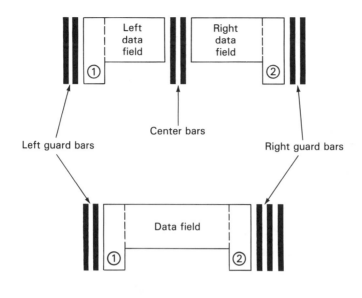

① Number system character

② Check digit

Figure 7-19   Universal Product Code (UPC) data string format.

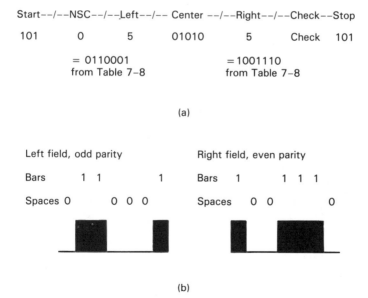

Figure 7-20   Layout of UPC bar code data string (a), left and right field data character (b).

5012 3452

Figure 7-21 European Article Numbering (EAN) bar code.

DOT MATRIX PRINTERS
- Paper cost is low.
- Medium to low density labels can be printed.
- Offers full form generation capability.
- Equipment available over wide cost range.
- Print quality may vary greatly.

FULL CHARACTER IMPACT PRINTING
- Paper cost is low.
- Equipment cost is high.
- High density labels are possible.
- Labels may be laminated.

LASER GENERATED LABELS
- Equipment cost is high.
- Will print on a wide range of substrates.
- High density labels possible.
- Techniques are not time proven.

REFERENCES

1. *Uniform Symbol Specifications (USS).* The Automatic Symbology Committee, Automatic Identification Manufacturers, Inc., Pittsburgh, Pa. 15238
2. *Bar Code Scanning,* MSI Data Corporation, Costa Mesa, CA 92626
3. *Bar Code Systems and Services,* Digital Data Services, Inc.
4. "Bar Codes." *Forbes,* 134 (1984), pp. 12–13.
5. *Fortune,* 106 (1982), pp. 98–101.
6. Maurer, Allen, *Laser: Light Wave of the Future,* New York: ARCO Publishing.
7. *UPC,* Uniform Product Code Council, Inc. 7051 Corporate Way, Suite 201, Dayton, OH 45459-4294
8. *Code 3 of 9 Specifications,* Military Standard 1189 Naval Publication and Forms Center, 5801 Tabor Ave., Philadelphia, PA 19120
9. *EAN,* International Article Numbering Association E.A.N., Rue Des Colonies, 54, Kolonienstraat, Bruxelles 1000 Brussel

EXERCISES

1. Identify
   (a) Five major bar codes in use.
   (b) The major types of bar code readers.
   (c) Can any bar code reader be used with any bar code? Explain your answer.
2. What code would you recommend to encode the following:
   (a) Part number 2A7B:
   (b) Part number 2473:
   (c) Part on nuclear sub:
   (d) Part for Buick car:
   (e) Box of cereal:
   (f) Part for European market:
   Explain the reasons for your recommendation. Could you have recommended more than one code for any of the items?
3. Draw the bar code data field for the number 172 in 3 of 9 format.
   Draw the bar code data field for the number 172 in 2 of 5 interleaved format.
4. Determine the difference of the length of the bar code symbol of the data fields for number 172.
   Does the difference depend on the number being recorded?
5. Recommend a system to encode part number 17253 for each of the following companies. Give reasons for your recommendation.
   (a) For General Dynamics on the nuclear submarines.
   (b) For General Motors on auto parts.
   (c) For General Foods on cereal.
6. Decode the bar codes in Figure 7-22.
7. Match the columns.

(a)

(b)

Figure 7-22   Horizontal (a) and vertical (b) 2 of 5 bar codes.

1. Bar coding _____ a. Light emitting diodes (LED), Infra-red (IR), coherent (Laser).

2. Codabar _____ b. Numeric system which is similar to UPC, but contains additional information.

3. Universal Product Code _____ c. Alpha-numeric system adopted for use by Department of Defense. Each character contains 3 wide elements out of a total of 9 elements.

4. Types of readers _____ d. Numeric system which encodes pairs of numbers, one by arrangement of wide and narrow bars, and the other by arrangement of wide and narrow spaces.

5. Interleaved 2 of 5 _____ e. Numeric system used for consumer products in the United States.

6. Code 3 of 9 _____ f. Encodes numeric and ten special characters; each data element contains 2 or 3 wide elements out of a total of 7 elements.

7. European article numbering _____ g. List of characters used by the system.

8. Character set _____ h. Unit which utilizes coherent radiation to read bar codes at a distance: contains light source and mechanism to move beam across the bar code.

9. Laser scanner _____ i. Automatic cost-effective method of identifying and acquiring product information.

GLOSSARY

ambient light: Light which is present in the environment around a machine vision system and generated from outside sources.

analog signal: The representation of signal by a smooth, continuous function curve of voltage vs time.

area diode array: A solid state video detector that consists of rows and columns of light-sensitive semiconductors. Sometimes referred to as a matrix array.

back lighting: The use of a light source placed behind an object so that a clear silhoutte of an object is formed. This is used when surface features on an object are not important.

bandwidth: The information carrying capacity of a circuit often expressed in frequency (MHz). As the bandwidth of the circuit increases, the information capacity increases.

binary image: A black and white digitized image represented as zeros and ones.

binary system: A vision system which creates a digitized image of an object in which each pixel can have one of only two values, such as black or white, one or zero.

boundary: The line formed by the joining of two image regions, each region having a different light intensity or pixel value.

155

CCIR: European monochrome broadcasting system standard which requires 625 lines at a 50-Hz frame rate.

cellular analysis: A procedure of assigning gray level and location information to each pixel and treating that unit of information as a cell.

centroid: The center of mass of an object having a constant density or the center of a given area or region in the case of an image.

CID, CCD, CPD: Types of semiconductor material: Charge Injection Device, Charge Coupled Device, and Charge Particle Device.

cluster analysis: A procedure of handling adjacent pixels within a given gray level as a unit or cluster.

c-mount: Type of lens mount conventionally found on many machine vision camera. Some cameras have U or bayonet mounts and use an adapter for C-mount lens.

contrast: The difference in light intensities between two regions in an image. This term is generally used to measure the difference between the lightest and darkest portion of an image.

convolution: Superimposing an $M \times N$ operator over an $M \times N$ pixel area (window) in the image, multiplying corresponding points together and summing the result. A normalization factor is frequently included in the process.

correlation: A correspondence between attributes in an image and a reference image.

detector: The light sensing portion of the electro-optical system. The detector translates light energy into an electrical signal.

digital image: A representation of a visual image by an array of brightness values.

digital to analog (converter): An electronic hardware device which converts a digital signal into a continuous voltage or current proportional to the digital input.

edge: A change in pixel values (exceeding some threshold) between two regions of relatively uniform values. Edges correspond to changes in brightness which can correspond to a discontinuity in surface orientation, surface reflectance, or illumination.

edge detection: The ability to determine the true edge of an object.

features: Simple image data attributes such as pixel amplitudes, edge or point locations and textural descriptors, or somewhat more elaborate image patterns such as boundaries and regions.

fiber optics: A technology in which light is used to transport information from one point to another.

frame: Digital data representing a single image at a specific point in time.

frame grabbing: Taking a set of data from a camera output representing a scanned frame and storing it in temporary memory for further analysis.

front lighting: The use of illumination in front of an object by observing the pattern made by an object intersecting the structured light.

gray level: A quantized measurement of image irradiance (brightness), or other pixel property usually given in integer values.

gray scale image: An image consisting of an array of pixels where each pixel has a value representing the average light intensity on the area. Typically, 16, 64, or 256 levels are possible for each pixel, depending on the number of bits available to process and store data.

HDTV: Huge density television which should be operational around 1992. It contains 1125 lines per 1/30 second on higher width to width ration than RS-170. Standard have not been finalized.

heuristics: "Rules of thumb," knowledge, or other techniques used to help guide a problem solution.

histogram: Frequency counts of the occurrence of each intensity (gray level) in an image usually plotted as the number of pixels with a given gray value vs gray level.

hough transform: A globel parallel method for finding straight or curved lines, in which all points on a particular curve map into a single location in the transform space.

hueckel operator: A method for finding edges in an image by fitting an intensity surface to the neighborhood of each pixel and selecting surface gradients above a chosen threshold value.

image: A projection of a scene into a plane. Usually represented as an array of brightness values.

image enhancement: The use of processing techniques to accentuate certain properties to improve the nature of the information received from an image.

image processing: Transformation of an initial image into a second image with more desirable properties, such as increased sharpness, less noise, and reduced geometric distortion.

intensity: The relative brightness of an image or portion of an image.

intrinsic characteristics: Properties inherent to the object, such as surface reflectance, orientation, incident illumination, and range.

isotropic: Rotation invariant, applying the operator and rotating the output will give the same results as rotating the input and applying the operator.

laplacian operator: The sum of the second derivatives of the image intensity in the x and y directions is called the Laplacian. The Laplacian operator is used to enhance value of pixel values adjacent to edge elements.

light: Electromagnetic radiation like radio waves except at a much higher (on the order of one thousand times) frequencies. The visible part of the spectrum is a small portion of the electromagnetic spectrum.

light emitting diode (LED): A semiconductor light source that emits visible light or infrared radiation.

linear array: A solid state video detector consisting of a single row of light sensitive semiconductor devices.

machine vision: The ability of an automated system to perform certain tasks normally associated with human vision, including sensing, image formation, image analysis, and image interpretation or decision making.

matrix array camera: A solid state camera which generates the image in the form of an $M \times N$ array of pixels.

morphology: Relationships between elements used in artificial intelligence approach to arrive at decision.

NTSC: Standard for color images in America which augments the RS-170 standard by utilizing a 3.58 MH color subcarrier on the video signal.

PAL: Standard for color images in the European broadcast system which is the equivalent of NTSC in the American system.

photodiode: A diode designed to produce photo current by absorbing light.

pixel: The smallest element of a scene, a picture element, over which an average brightness value is determined and used to represent that portion of the scene. Pixels are arranged in a rectangular array to form a complete image of the scene.

pixel value: Average brightness value over the pixel area; usually rounded off to integer value.

processing speed: The time required for a vision system to analyze and interpret an image.

reflectance: The ratio of total reflected to total incident illumination at each point.

region: A set of connected pixels that show a common property such as the small gray level, color or texture, in an image.

region growing: Process of initially partitioning an image into elementary regions with a common property (such as gray level) and then successively merging adjacent regions with small differences in the selected property, until only regions with large differences between them remain. Similar to clustering.

registration: Processing images to correct geometrical and intensity distortions, relative translational and rotational shifts, and magnification differences between one image and another or between an image and a reference map. When registered, there is a one-to-one correspondence between a set of points in the image and in the reference.

resolution: The smallest feature of an image which can be sensed by a vision system. Resolution is generally a function of the number of pixels in the image, with a greater number of pixels giving better resolution.

rotationally insensitive operator: The Laplacian operator, related to the magnitude of the derivative of the intensity gradient, is insensitive to the direction of a line and yields edge elements at pixel points where the Laplacian is zero. Thus, discrete approximations to the Laplacian have proved useful to line finding.

RS-343-A: Standard which covers timing relationships for images with 675, 729, 875, 945, and 1023 vertical lines.

RS-232C: Standard computer interface data link used by CRT and TTY terminals.

RS 170: Standard TV monitor interface; 525 scans, interlaced, 1/30 second, active time 52 microseconds, retrace time 11.6 microseconds.

run-length encoding: A data compression technique in which an image and only the lengths of runs of consecutive pixels with the same property are stored.

segmentation: The process of dividing a scene into a number of individually defined regions, or segments.

signal to noise ratio (SNR): The ratio between the usable signal and any extraneous noise signal present. This ratio is expressed in dB. If the SNR is exceeded, the transmitted signal quality will be unacceptable.

stereoscopic approach: Use of triangulation between two or more views, obtained from different positions, to determine range or depth.

strobe light: An electronic flash tube which produces a short burst or high intensity flash of microseconds duration. It is used to freeze images of moving objects for vision systems.

structured lighting: Sheets of light and other projective light patterns used to determine shape and/or dimensions of an object by observing the pattern made by an object intersecting the structured light.

symbolic description: Nonimage scene descriptions such as graph representations.

template: An object outline to be matched with an observed image field. Usually performed at the pixel level.

texture: A local variation in pixel values that repeats in a regular or random way across a portion of an image or object.

thresholding: Separating elements or regions of an image for processing based on pixel values above or below a chosen (threshold) value or gray level.

triangulation: A method of determining distance by forming a right triangle consisting of a light source, camera and sample.

wavelength: The distance between two successive peaks of a signal.

windowing: A technique for reducing data processing requirements by electronically defining only a small portion of the image to be analyzed. All other parts of the image are ignored.

INDEX

K

"K" normalizing constant, 92

L

Label production, 148
Lag, cameras, 41
Laplacian operator, 113
Laser, 132
  scanner, 132
LED (Light Emitting Diode), 132
Lens mounting, 37
Light:
  ray path, 31
  spectrum, 9
Lighting, 21
Linear arrays, 39
Line detection, 110
Local operator, 86
Low-pass-filter, 106

M

Machine vision, 3
  inspection cost, 11
Magnification, 32
Matching patterns, 120
Matrix array camera, 47
Measurement error, 63
Micron, 7
Moments, 82
Monadic, 87
Morphologic, 123
MOS, 46
Multiplication, images, 96
Multi-point operators, 91

N

Neighborhood operator, 106
Newvicon, 39
Noise, 106
Non-uniform lighting, 57

O

Object plane, 32
Off-set structured lighting, 29

Operators, 87
Optics, 33
Output, 49

P

Parameter, image, 82
Pattern matching, 120
Perception, 10
Perimeter, 82
Photodiode, 39
Pixel, 54
  area, 54
  map, 55
  value, 54
Plumbicon, 39
Point-by-point transform, 86
Polar coordinates, 121
Preprinted labels, 148
Processing, 47
Production, labels, 148

Q

Quality of measurements, 6
Quantification error, 62

R

Radius of curvature, 32
Reader aperture, 135
Regional operator, 49
Registration, 92
Resolution, 135
  index, 135
Response speed, 7
RGB color system, 68
Roberts operator, 116
RS-170 format standard, 80
Rounding off, 62

S

Sampling, 42
  rate, 43
Saticon tube, 39
Scanner specification, 133
Scanning, 40